Real-Time Structural Health Monitoring of Vibrating Systems

Basuraj Bhowmik
School of Mechanical & Materials Engineering
University College Dublin, Dublin, Ireland
and
Department of Civil Engineering
Indian Institute of Technology, Guwahati, India

Budhaditya Hazra
Department of Civil Engineering
Indian Institute of Technology, Guwahati, India

Vikram Pakrashi
School of Mechanical and Materials Engineering
University College Dublin, Dublin, Republic of Ireland

CRC Press
Taylor & Francis Group
Boca Raton London New York

CRC Press is an imprint of the
Taylor & Francis Group, an **informa** business
A SCIENCE PUBLISHERS BOOK

First edition published 2022
by CRC Press
6000 Broken Sound Parkway NW, Suite 300, Boca Raton, FL 33487-2742

and by CRC Press
4 Park Square, Milton Park, Abingdon, Oxon, OX14 4RN

CRC Press is an imprint of Taylor & Francis Group, LLC

Library of Congress Cataloging-in-Publication Data (applied for)

ISBN: 978-0-367-35095-6 (hbk)
ISBN: 978-1-032-16953-8 (pbk)
ISBN: 978-0-429-35134-1 (ebk)

DOI: 10.1201/9780429351341

Typeset in Times New Roman
by Radiant Productions

Preface

Detection of features of interest in dynamical systems are becoming increasingly important. While these systems vary widely, this book mainly considers Mechanical and Civil engineering sectors with cross-cutting overall themes around our built infrastructure. The term *detection* is also interpreted quite widely. It can refer to a detection of anomaly or a sudden change in terms of its presence and position in time. More extensively, it can also raise questions around classification of such anomalies, estimation of remaining performance or capacity of such systems, or the estimation of parameters of structural health (e.g., stiffness), which can make the problems more complex. The features of interest may not necessarily be related to sudden or gradual damage, but can also be related to parameters representing their performance with respect to control, repair or change in other operational conditions. Detection of features of interest in real-time, using only the output dynamic responses can be particularly relevant for these systems since we typically tend to have relatively good control over the output responses, and the extensive rise of sensors and data has made it possible to implement such solutions. Historically, this was not the case and the current rise in sensing, computing and analysis capabilities create a new paradigm in real-time detection. With the rise of model updating and digital twinning, such real-time implementations become even more relevant.

Despite the promises of real-time detection, there are a number of challenges that exist. The need for real-time monitoring and the interpretation of results change from problem to problem. Additionally, there exists inadequate evidence base – both numerically and experimentally – to understand, quantify and implement such solutions for different sectors. While the overall topic of Structural Health Monitoring has expanded quite significantly, there is a need to investigate the real-time aspects of such detection, their performance and limitations to those performance further. There are variabilities and uncertainties to the inputs and the system – along with measurement errors. Operational variabilities are also a challenge as are the various types and extents of nonlinearities in a measured system, along with their interaction with stochasticity. These variabilities can often overshadow detection.

This book attempts to address many such problems and provide an overall structure around the real-time detection paradigm. There is an increasing move away from trying to seek the 'best method' for a detection to a 'set of admissible methods' that work adequately in relation to the need of detection of a certain sector. This book has tried to be in line with that movement, while providing numerical and experimental evidence for comparison. We hope that the book will be relevant for researchers and practitioners alike to provide a real-time approach towards monitoring and will encourage further research in this field.

Acknowledgements

The need for real-time monitoring of features of interest in various systems is becoming more relevant and accessible with time. We sincerely hope that this book helps those who are interested in such monitoring, or would like to see if real-time monitoring could interest them. We also hope that this book continues to support the interest and evolution in real-time monitoring in times to come. The book would not have been possible without the help of several people and we are thankful to all of them.

The Covid19 pandemic has taken over our lives, in some shape or form during most of the duration spent in writing this book. It is important to mention here the collective resilience with which we continue to endure this challenging time and continue to persevere.

Basuraj Bhowmik

I would like to thank my co-authors for their support, insightful comments, and valuable suggestions. Thank you to all the people I have met in my trips over the years – some of whom I kept in touch with, some of whom I worked with, and some who changed the direction of my life. I would not have been here today writing these words. Thanks to my readers – researchers, practitioners, and students – who believed in knowing, learning, and coming off to a trip with me. Thank you to Anwesha: You are the one I take my trips with now. A final thanks to my parents for making this wild and uncertain trip called 'life', easy.

Basuraj Bhowmik dedicates this book to the love of his life, Anwesha, for comforting, consoling, and contemplating all the decisions he ever took. "You are my idea of beautiful – no one else ever made sense."

Budhaditya Hazra

I wish to gratefully acknowledge the support of my family, friends and students. This work has crystallized from the untiring efforts of some very diligent students like Tapas Tripura, Manu Krishnan, Ankush Gogoi and Satyam Panda. Finally, I can't thank my cute little daughter Ayushi Hazra and my dear wife Mitali Hazra enough for everything that they have brought into my life.

Budhaditya Hazra would like to dedicate the book to his wife, Mitali, for all her pains and sacrifices endured to make our lives a meaningful and a savoring experience.

Vikram Pakrashi

I would like to acknowledge the support of my team (Dynamical Systems and Risk Laboratory, UCD Centre for Mechanics), my co-authors and collaborators around the world who made this book possible. Thanks to the funding of many agencies around this topic throughout the years, where Science Foundation Ireland (through MaREI Centre), Irish Research Council and Sustainable Energy Authority of Ireland (SEAI) feature prominently. I would also like to acknowledge the support and scientific criticism of Prof. Bidisha Ghosh, Trinity College Dublin, Ireland and the love of Naloke Ghosh, which made me continue with this work.

Vikram Pakrashi would like to dedicate this book to all researchers who made this possible.

17/06/2021 | Dublin | Ireland

Contents

Symbols

s	Seconds
$p(x(t_1), x(t_2); t_1, t_2)$	Second order probability density function
$\mu_x(t)$	First order statistics – mean
$R_{xx}(t_1, t_2)$	Second order statistics – variance
$\bar{y}$	Temporal statistics – mean for stationary signal y
y_{rms}	Root mean square of signal y
a_0, a_n, and b_n	Coefficients of a Fourier series
ω_n	Frequency of a harmonic
$X(\omega)$	Fourier transform of the signal $X(t)$
$\mathbf{M}$	Mass matrix
$\mathbf{K}$	Stiffness matrix
$\mathbf{C}$	Damping frequency
$\mathbf{H}(\omega)$	Frequency response function
$S_x(\omega)$	Power spectral density function
$R_X(\tau)$	Auto correlation function
$\delta(\tau)$	Dirac delta function
$\mathbf{C_0}$	Covariance matrix corresponding to initial monitoring phase
$\mathbf{V_0}$	Eigen vector matrix corresponding to initial monitoring phase
$\mathbf{\Lambda_0}$	Eigen value matrix corresponding to initial monitoring phase
$\mathbf{C_n}$	Covariance matrix corresponding to real-time monitoring phase

$\mathbf{V}_n$	Eigen vector matrix corresponding to real-time monitoring phase
$\mathbf{\Lambda}_n$	Eigen value matrix corresponding to real-time monitoring phase
$\delta\mathbf{C}$	First order eigen perturbation of the covariance matrix
$\tilde{\mathrm{Y}}$	Augmented trajectory matrix
$\mathbf{V}_j$	Individual eigenmatrix of jth channel
$\mathcal{P}$	Principal orthogonal component
$\mathcal{Q}$	Linear normal coordinates
$\mathcal{E}$	Error term

PART I

Chapter 1

Introduction

In the past two decades of accelerated technological development, the reliance on structural and mechanical systems such as aircrafts, buildings, bridges, railways, power generation systems and defence systems is rising steadily. The need for structural health monitoring (SHM) can be traced to the earliest endeavours of mankind to conceptualize, understand and worry about deterioration, with attempts at prolonging the service life or repairing of a structure. This is a natural response to the fact that built infrastructures deteriorate over time – due to a gradual loss of pre-stress, stiffness and tension – which makes it essential for asset managers to arrive at informed decision regarding the optimality of repair. A significant portion of the built infrastructure could benefit from extensive condition monitoring and data-driven, informed maintenance. However, constraints such as improper planning, negligence in management and unavailability of adequate resources impede their systematic implementation. In terms of the design and introduction of new engineering systems, low safety margins are sometimes adopted to develop cost-effective designs. Therefore, the demand for detection of damage at the earliest possible time becomes crucial in order to prevent failures leading to life-safety and socio-economic losses.

Structural health monitoring (SHM) is usually defined as the process of implementing a conventional inspection and damage assessment using non-destructive testing (NDT) and statistical characterization to identify defects in structural response. SHM can be considered to be a process of implementing damage detection strategies to arrive at a data-driven decision regarding the health of the system. SHM often involves long-term monitoring of a dynamical system using periodically spaced measurements. Dense sensor arrays are becoming popular

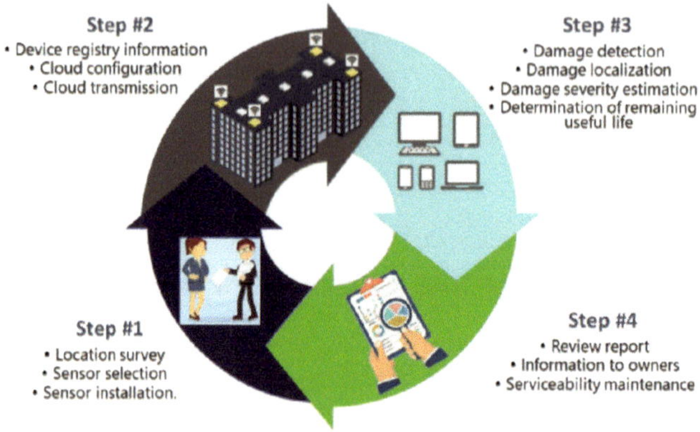

Figure 1.1: Components of a SHM system.

for measurement and they are connected to the mechanical system that provides features of interest, often upon excitation of a dynamic force. Ambient vibrations or free vibration data are also used sometimes. Subsequent transmission of data (usually over a registered cloud infrastructure) and its analysis (offline or online) provide damage sensitive features (DSFs) that determine the current state of system health. An outline of a complete SHM process is illustrated in Fig. 1.1.

According to an estimate, the global market for SHM is estimated to grow from USD 1.5 billion (in 2018) in excess of USD 3.4 billion by 2023 – rising at a compound annual growth rate (CAGR) of 17.9% [2] during the forecast period. With 9.1% structurally deficient bridges in 2016, the Americas the need for SHM is expected to increase, especially due to increase in fast-growing infrastructural developments, increasing investment and improvement needs of the ever-ageing built infrastructure [1]. In line with these projections, economies such as China, Japan, South Korea and India are expected to generate revenue opportunities for SHM providers in the Asia–Pacific (APAC) region as well [2]. Several factors like public safety, preservation of heritage structures and investment in the public sector, shape the stronghold of SHM market in the APAC region, which is expected to witness significant CAGR through 2030. EU already has embraced the idea of SHM in several sectors and while a number of companies exist engaging with the process, there is a need for creating further guidelines, recommendations and normative documents, along with extensive numerical and experimental evidence bases. This is expected to create a strong demand for SHM systems, which will facilitate informed decision-making processes and reduce the overall maintenance of infrastructure [2].[1]

[1]At the time of writing this book, the 2017 ASCE Infrastructure report card provided the most updated statistics on the global SHM market.

1.1 Need for SHM development

The implementation of monitoring for industrial products is motivated by their socio-economic impact, safety and operational concerns. The need for condition assessment, therefore, becomes increasingly relevant for publicly and privately funded industries, in general – and even more crucial for government organizations in particular, where taxpayers' money is involved. As an example, in the context of wind power generation, anomalies (usually referred to as 'downtimes') caused by a variety of off-nominal conditions, result in failures and costly repairs. The importance of early damage detection here translates to data-driven actions that result in significant energy benefits and cost savings for the industry. In addition to preventing further damages, anomaly detection provides institutional awareness to asset managers regarding the source of abnormal behaviour and its severity for developing a mitigation plan. For built infrastructure, the emphasis is focussed more towards providing a means of minimizing uncertainty associated with damage assessments. There is clearly a twofold aspect of SHM from the civil engineering context. For example, post-seismic events lead to infrastructure owners assessing the remaining serviceability and prompt reoccupation of buildings – in an attempt at significantly mitigating economic losses arising due to earthquakes. SHM not only provides information about the present state of the system's health but also assesses if the structure needs to be shut down to prevent accidents or remain functional with a specified level of reliability. Second, even though an increasing number of built infrastructures are ageing or exceeding design life, these are being used to counteract economic losses. Therefore, implementing a strategic monitoring system becomes essential in such cases. With the rise of renewable energy, maintenance for onshore and offshore turbines have also become another important aspect in recent times.

Irrespective of the sector of implementation, an SHM module should have the functionality to acquire, analyse and validate structural response data on the basis of which decisions related to continuity of service can be taken. With the advent of computationally efficient methods, low-cost sensor development and improved data acquisition have progressed leaps and bounds. This has enhanced the outlook on damage detection and its characterization, its analysis and interrogation – realizing the fact that SHM should not only be considered as a tool for maintenance, but as a yardstick on which a decision-informed risk-based methodology could be potentially framed.

Among all the important aspects of building management, the most relevant one is *maintenance*. Maintenance philosophies strive to attain minimum unplanned downtime – irrespective of the nature and the behaviour of the dynamical systems – to address critical situations that could potentially take down operations. Traditional *reactive* approaches have now been replaced with sophisticated *predictive* building maintenance strategies. While a reactive *run-to-failure* strategy focusses mainly on restoring assets back to normal operation after a breakdown, it ensures low maintenance costs when the asset is utilized to its maximum potential. However, for physical infrastructure such as bridges and buildings, the functionality and efficiency in performance are the major concerns for which continuous monitoring is essential. The concepts of SHM are firmly embedded into *predictive maintenance* strategies, ensuring that assets are shut down before an inevitable failure – based on data collected by methodical acquisition and analysed by algorithms that determine trends to failure. This contributes to optimizing asset efficiency, reduction of energy costs and most importantly, extends the serviceability of the assets by reducing the need of costly repairs. Table 1.1 summarizes the key concepts of the discussed maintenance strategies.

Table 1.1: Key differences between maintenance strategies.

Reactive Maintenance	**Predictive Maintenance**
Run-to-failure strategy	Predict and prevent (failure) strategy
Repair or replace faulty components *once*	Extend serviceability of assets
Impaired productivity due to unplanned downtime	Scheduled maintenance leading to planned work
No maintenance carried out before failure	Continuous monitoring
Manual replacement and repair	Automated real-time SHM strategy

1.2 Definitions

1. **Damage**: A deteriorating phenomenon that induces a change in the physical, material or geometrical properties of a structure is known as 'damage' [3–6]. An easier interpretation of damage is the process that leads to the deviation of system health from its *sound state*. In the context of built infrastructures, damage is defined as the change in system parameters such as mass, stiffness and damping, that may compromise structural integrity and affect its serviceability [7–11]. In terms of modal properties, changes in natural frequencies [10, 24], mode shapes [7, 11, 14, 23] and damping ratios [9, 18] have been consistently considered as effective indicators of damage. In the pioneering work penned by Farrar and Worden, *damage* has been aptly defined as – 'intentional or unintentional changes to the material and/or geometric properties of mechanical systems, including changes to the boundary conditions and system connectivity, which adversely affect the current or future performance of these systems' [17].

2. **Damage detection**: A process involving data acquisition, analysis and its inference on the change of structural properties of a dynamical system. This is carried out in the context of closely related disciplines such as

condition based monitoring (CBM), non-destructive testing and evaluation (NDT&E) and damage prognosis [15, 17].

3. **Condition based monitoring (CBM)**: The process of monitoring a pre-selected set of parameters affecting the condition (or performance) of a machinery. These parameters (vibration, temperature, among others) are indicative of an impending fault and forms a major part of predictive maintenance strategy [10, 15]. Although analogous to SHM, CBM in particular, addresses damage detection in industrial and rotating systems. Both SHM and CBM have the potential to be applied in *real-time*, i.e., *in-situ* analysis of streaming data during the operation of the system of interest. Contrary to this aspect lies the overarching notion of SHM that embraces monitoring of both built infrastructure and industrial machinery, in tandem. SHM extends its applicability in providing damage information for structures under extreme events (earthquake or blasts) and amidst operational and environmental variability.

4. **Non-destructive testing and evaluation (NDT&E)**: NDT is primarily used for damage characterisation for estimating the severity of the induced damage. This is usually carried out offline, when the location of the potential damage is known apriori [15, 16]. NDT&E uses common physical techniques of investigation – ultrasonics, radiography, eddy current testing, magnetic particle inspection and dye penetrants – from which ultrasonics and radiography are most commonly used techniques for NDT of composites and bonded joints [15]. Although similar in functioning, the distinct differences between NDT and SHM can be summarized as:

    (a) NDT provides state assessment at a particular time stamp without necessarily addressing the effect or extent of deterioration, which would otherwise be easily explained through stochastic SHM processes [4, 17].

    (b) The notion of prognosis (estimation of remaining useful life of the structure) is embedded in the auspices of SHM which requires deep knowledge of material failure mechanisms. A subsequent NDT can be further triggered based on predefined thresholds set by a SHM module.

    (c) For a complex integrated system such as a bridge, it seems impossible to design and allocate individual NDT modules. Instead, SHM is more likely to incorporate the functionalities of individual NDT systems and establish a damage threshold to identify zones where inspection is required [19].

5. **Damage prognosis (DP)**: The accurate estimation of a system's remaining useful life (RUL) under fatigue loading, continuous degradation and anomalies is known as damage prognosis (DP) [21]. Emerging techniques amalgamating SHM and DP have addressed challenges arising due to the complexity and uncertainty in service environments and multidisciplinary damage mechanisms. A robust DP strategy relies on the knowledge of the present status of system components monitored by a SHM module [20–22].

1.3 Components of SHM

Advances in improved sensing, low-cost data transmission, and analysis have alleviated challenges associated with implementation of SHM systems. Big-data analytics, data-mining, ultra-low power cloud infrastructure and machine learning have spearheaded SHM into an established, proven technique for better management of built infrastructure. However, shortcomings in fundamental knowledge, technology demands, lack of skilled personnel for routine maintenance and uncertainties due to *complexities of reality* still need to be carefully addressed for a meaningful implementation of SHM [23]. Evidently, augmentation of statistical models for differentiating features from undamaged and damaged states of a system, is an evolving area – which is *nascent* at worst to *developing* at best. The functional relationships between extracted features of interest and the state of a monitored structure is often determined by the use of *machine learning* (ML) paradigms.

In situations where prior knowledge about the pristine state of a system is known, the data acquired from the monitoring state can be referenced to draw conclusions regarding the extent of damage. The algorithms employed for straightforward comparison between two or more states of a system falls under the *supervised learning* category. Regression analysis [27, 28] and group classification theories [25, 26] form the key methods under this category. These algorithms are mostly *offline* and *reference-based*, owing to the initial acquisition of data and its comparison to a healthy state. In the context of SHM, *unsupervised learning* is employed when the data from only the monitored phase is known and the decision regarding the state of the system has to be gauged without prior reference to its healthy state. *Real-time* novelty detection methods are the key algorithms used in this situation [9, 10]. These methods detect the instant of damage on the basis of the information acquired in the previous instant of time. Recent advances in the field of first order eigenperturbation (FOEP) techniques have proven to be computationally effective and robust at detecting online damage for simulated case studies, experimental test beds and real-life systems [11–13].

An effective SHM design constitutes a hierarchical institution discussed in [24]. The answers to the following questions form the preliminary components of any SHM module [17, 24]:

1. Existence: Is there damage in the system?
2. Location: If yes, then where is the damage in the system?
3. Type: What kind of damage is present?
4. Extent: How severe is the damage?
5. Prognosis: How much is the remaining useful life?

The sequential answers to the above questions provide a thorough understanding of the state – damaged or undamaged. Detection studies using the supervised learning model can be employed in conjunction with analytical models and statistical procedures, which can – in theory – be used to reference the baseline with the monitored condition, to infer damage. Attribution to an online detection scheme can be effectively attained using unsupervised learning schemes. These methods are beneficial in identifying damage patterns using statistical paradigms and alleviates the necessity of employing a numerical model. Figure 1.2 summarizes the key constituents of a SHM design.

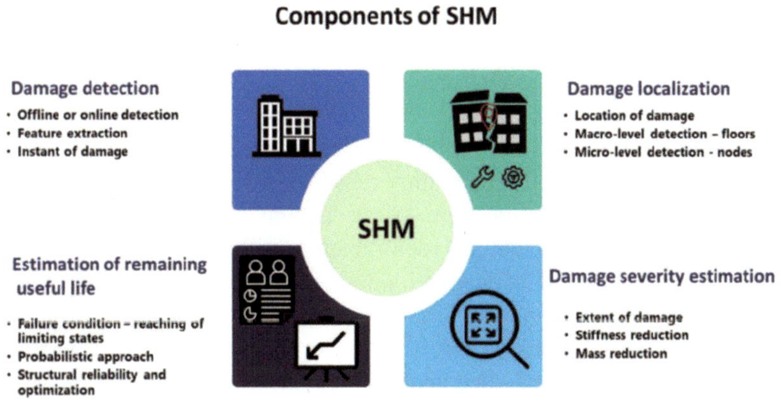

Figure 1.2: Components of a complete SHM module.

1.4 Need for real-time SHM

Infrastructure owners, stakeholders (asset managers) and end-users (residents) often require knowledge about the integrity and reliability of the structural system and its components in *real-time* to execute preventive actions. This enables timely and early decisions on whether component serviceability has been impaired or if it can still function with a certain level of reliability. The use of real-time detection enables immediate updating of risk by providing the instant, location, and severity of the damage. Recursive analysis of system health has the following characteristics:

1. Value of information is allocated by linking design, construction and service-life maintenance together that creates a provision for development of normative documents, including codes of practice.

2. RUL thus obtained is calculated based on updated probability measures from continuous assessments and recursive SHM strategies. This creates an evidence-base around the functioning of the structural components – which would otherwise have been evaluated based on high global safety factors, leading to unnecessary maintenance and costly repairs.

Damage, if not detected in time, can have a significant impact on the functionality, integrity and serviceability of the system as a whole, which could lead to severe economic losses and adversely affect public safety. Visual inspections are the basic and most common maintenance and detection technique used for a long period of time, but this traditional approach can only be applied for simple structures. In the case of complex structures, visual inspections may not be possible due to restricted accessibility. To alleviate this drawback, a significant number of damage detection algorithms and strategies have been proposed in recent times. Changes in modal properties of structures during their service life are strongly related to impending damage, which makes accurate estimation of modal properties an essential step in SHM. Therefore, both monitoring and accurate identification of real-time modal properties (modal frequency, damping ratio, and mode shape) are of crucial importance in order to have a good estimate in SHM. However, the evolution of real time damage detection schemes capable of conducting baseline-free damage identification, still poses a formidable challenge, primarily, due to the underdevelopment of algorithms that are amenable towards real time implementation. While large data streams might be impressive, the ability of an online framework to validate structural damage in real-time, is important. Statistical learning, data analytics and the development of subsequent numerical and experimental evidence base can make such a real-time detection method robust and efficient. As the occurrence of damage is often a sudden event, it requires the damage sensitive features (DSFs) to function online, in a recursive fashion, for continuously streaming data. This can be carried out extensively by a family of recursive damage detection techniques called first order eigenperturbation (FOEP) methods and tailoring them towards real time SHM [29–33]. This will be discussed in detail in the subsequent chapters.

1.5 Motivation for this book

The emerging interest for SHM has initiated an internationally acclaimed conference series worldwide. Recent conferences [40–44] have ensured that the dissemination of quality research on the topic of SHM is globalized and accessible to researchers and practitioners – especially in terms of conference proceedings and special journal issues [44, 45]. These conferences have demonstrated the growing interest of seasoned researchers and novice learners who are extremely passionate to pursue this exciting and continuously evolving field of study.[2] The emergence of specialized courses on SHM technologies and methodologies have sparked widespread interest not only among academicians but industrial partners who intend to develop the hardware aspect of monitoring. The inaugural edition of the first refereed journal dedicated solely to SHM and its applications was initiated in 2002 [37]. Since its inception, researchers and experts worldwide have made significant efforts in initiating similar journals tending to the applications of SHM – notably [38] and [39] – involving structural control and SHM of built infrastructures. However, the limited number of published books on SHM [16, 17, 45, 46]– in sharp contrast to the commendable conference efforts – have led the authors to firmly realize the need for establishing a comprehensive work on the topic of *real-time* monitoring of vibrating systems. This book is therefore, an attempt at providing the engineering community an updated outlook on vibration-based *recursive SHM* strategies – aspects of which are arguably neglected in the usual coverage of SHM to date.

1.6 Organization of the book

Much of the material for this book strongly relies on pedagogical practices of mathematics – especially linear algebra and matrix transformations – that build the foundations of this book. **Chapter 2** discusses the fundamentals of mathematical operations required to venture through this book.

Single-sensor real-time damage detection technique forms the highlight of **Chapter 3**. Cost considerations, improper accessibility, and unavailability of good quality sensors pose threat to monitoring of infrastructure which can be resolved by optimally placing a single sensor and extracting good quality data.

Chapters 3 to **6** cover details of multi-sensor real-time damage detection. Recently emerging real-time formulations have been discussed that provide concepts required to understand online damage detection applications. Spanning across these chapters is a unified treatment of the topics and the authors have

[2] At the time of writing this book, the world is still reeling from the outbreak of the COVID-19 pandemic. Most of the strategic conventions such as the EWSHM [34], ASCE-EMI [36] and EMI-PMC [35] conferences have been postponed till 2021.

ensured that sign conventions, symbolisms, and pedagogy are maintained and consistent throughout.

Chapter 7 discusses real-time modal identification for vibrating systems which is not usually achieved in conventional textbooks. The contents of this chapter treats the methods previously discussed within a mathematically rigorous framework and provides the fundamental understanding of eigenspace calculation – leading to the evolution of vibratory modes – in the transformed or reduced order domain.

Finally, **Chapter 8** provides an executive overview of all the practical applications that have so far been carried out using the methods. Specific applications not only include the stand-alone algorithms for real-time damage detection, but also involve field applications towards online modal identification.

This book can be used for teaching, research, and practice. It not only provides the fundamental understanding of real-time structural health monitoring, but also instils confidence among the readers to practice and implement in their research work. Online SHM is going to be more popular over time in traditional civil engineering and its alignment with statistical learning approaches make this the forerunner contributor in the emerging field of *in-situ* monitoring.

References

[1] Grigg, N. S. (2015). Infrastructure report card: Purpose and results. Journal of Infrastructure Systems, 21(4): 02514001.

[2] Structural Health Monitoring Market. (2018). MarketsAndMarkets, https://www.marketsandmarkets.com/Market-Reports/structural-health-monitoring-market-101431220.html.

[3] Farrar, C. R. and Worden, K. (2007). An introduction to structural health monitoring. Philosophical Transactions of the Royal Society A: Mathematical, Physical and Engineering Sciences, 365(1851): 303–315.

[4] Balageas, D., Fritzen, C. P. and Güemes, A. (eds.). (2010). Structural Health Monitoring (Vol. 90). John Wiley & Sons.

[5] Farrar, C. R. and Worden, K. (2012). Structural Health Monitoring: A Machine Learning Perspective. John Wiley & Sons.

[6] Doebling, S. W., Farrar, C. R. and Prime, M. B. (1998). A summary review of vibration-based damage identification methods. Shock and Vibration Digest, 30(2): 91–105.

[7] Gul, M. and Catbas, F. N. (2009). Statistical pattern recognition for structural health monitoring using time series modeling: Theory and experimental verifications. Mechanical Systems and Signal Processing, 23(7): 2192–2204.

[8] Balsamo, L. and Betti, R. (2015). Data-based structural health monitoring using small training data sets. Structural Control and Health Monitoring, 22(10): 1240–1264.

[9] Curadelli, R. O., Riera, J. D., Ambrosini, D. and Amani, M. G. (2008). Damage detection by means of structural damping identification. Engineering Structures, 30(12): 3497–3504.

[10] Worden, K. and Dulieu-Barton, J. M. (2004). An overview of intelligent fault detection in systems and structures. Structural Health Monitoring, 3(1): 85–98.

[11] Wahab, M. A. and De Roeck, G. (1999). Damage detection in bridges using modal curvatures: Application to a real damage scenario. Journal of Sound and Vibration, 226(2): 217–235.

[12] Pandey, A. K., Biswas, M. and Samman, M. M. (1991). Damage detection from changes in curvature mode shapes. Journal of Sound and Vibration, 145(2): 321–332.

[13] Hearn, G. and Testa, R. B. (1991). Modal analysis for damage detection in structures. Journal of Structural Engineering, 117(10): 3042–3063.

[14] Kim, J. T., Ryu, Y. S., Cho, H. M. and Stubbs, N. (2003). Damage identification in beam-type structures: Frequency-based method vs mode-shape-based method. Engineering Structures, 25(1): 57–67.

[15] Adams, R. D. and Cawley, P. D. R. D. (1988). A review of defect types and non-destructive testing techniques for composites and bonded joints. NDT International, 21(4): 208–222.

[16] Rose, J. L. (2004). Ultrasonic guided waves in structural health monitoring. In Key Engineering Materials (Vol. 270, pp. 14–21). Trans. Tech. Publications Ltd.

[17] Tripura, T., Bhowmik, B., Pakrashi, V. and Hazra, B. (2019). Real-time damage detection of degrading systems. Structural Health Monitoring, 1475921719861801.

[18] Ray, A. (2004). Stochastic measure of fatigue crack damage for health monitoring of ductile alloy structures. Structural Health Monitoring, 3(3): 245–263.

[19] Raghavan, A. (2007). Guided-wave Structural Health Monitoring (Doctoral Dissertation).

[20] Kulkarni, S. S. and Achenbach, J. D. (2008). Structural health monitoring and damage prognosis in fatigue. Structural Health Monitoring, 7(1): 37–49.

[21] Farrar, C. R. and Lieven, N. A. (2007). Damage prognosis: the future of structural health monitoring. Philosophical Transactions of the Royal Society A: Mathematical, Physical and Engineering Sciences, 365(1851): 623–632.

[22] Ling, Y. and Mahadevan, S. (2012). Integration of structural health monitoring and fatigue damage prognosis. Mechanical Systems and Signal Processing, 28: 89–104.

[23] Catbas, F. N., Ciloglu, S. K., Hasancebi, O. Ğ. U. Z. H. A. N., Grimmelsman, K. et al. (2007). Limitations in structural identification of large constructed structures. Journal of Structural Engineering, 133(8): 1051–1066.

[24] Rytter, A. (1993). Vibrational Based Inspection of Civil Engineering Structures (Doctoral Dissertation, Dept. of Building Technology and Structural Engineering, Aalborg University).

[25] Farrar, C. R., Duffey, T. A., Doebling, S. W. and Nix, D. A. (1999). A statistical pattern recognition paradigm for vibration-based structural health monitoring. Structural Health Monitoring, 2000: 764–773.

[26] Figueiredo, E., Park, G., Figueiras, J., Farrar, C. et al. (2009). Structural health monitoring algorithm comparisons using standard data sets (No. LA-14393). Los Alamos National Lab. (LANL), Los Alamos, NM (United States).

[27] Magalhães, F., Cunha, Á. and Caetano, E. (2012). Vibration based structural health monitoring of an arch bridge: From automated OMA to damage detection. Mechanical Systems and Signal Processing, 28: 212–228.

[28] Dervilis, N., Worden, K. and Cross, E. J. (2015). On robust regression analysis as a means of exploring environmental and operational conditions for SHM data. Journal of Sound and Vibration, 347: 279–296.

[29] Bhowmik, B., Tripura, T., Hazra, B. and Pakrashi, V. (2019). First-order eigen-perturbation techniques for real-time damage detection of vibrating systems: Theory and applications. Applied Mechanics Reviews, 71(6).

[30] Krishnan, M., Bhowmik, B., Hazra, B. and Pakrashi, V. (2018). Real time damage detection using recursive principal components and time varying auto-regressive modeling. Mechanical Systems and Signal Processing, 101: 549–574.

[31] Bhowmik, B., Krishnan, M., Hazra, B. and Pakrashi, V. (2019). Real-time unified single-and multi-channel structural damage detection using recursive singular spectrum analysis. Structural Health Monitoring, 18(2): 563–589.

[32] Bhowmik, B., Tripura, T., Hazra, B. and Pakrashi, V. (2020). Real time structural modal identification using recursive canonical correlation analysis and application towards online structural damage detection. Journal of Sound and Vibration, 468: 115101.

[33] Bhowmik, B., Tripura, T., Hazra, B. and Pakrashi, V. (2020). Robust linear and nonlinear structural damage detection using recursive canonical correlation analysis. Mechanical Systems and Signal Processing, 136: 106499.

[34] 10th European Workshop on Structural Health Monitoring. (2021). EWSHM, http://www.ewshm2020.com/.

[35] Engineering Mechanics Institute Conference and Probabilistic Mechanics and Reliability Conference. (2021). ASCE-EMI-PMC, https://www.emi-conference.org/.

[36] ASCE Engineering Mechanics Institute International Conference. (2021). https://sites.durham.ac.uk/emi2020-ic/.

[37] Journal of Structural Health Monitoring, SAGE Publications. https://journals.sagepub.com/home/shm.

[38] Journal of Structural control and Health Monitoring, Wiley Online Library. https://onlinelibrary.wiley.com/journal/15452263.

[39] Journal of Civil Structural Health Monitoring, Springer. https://www.springer.com/journal/13349.

[40] International Operational Modal Analysis Conference. (2019). http://iomac.eu/.

[41] The 12th International Workshop on Structural Health Monitoring. (2019). https://web.stanford.edu/group/sacl/workshop/IWSHM2019/.

[42] International Association for Bridge and Structural Engineering Congress. (2019). https://www.iabse.org/IABSE/Events/Newyork2019/IABSE/events/Conferences_files/Newyork2019/Home.aspx?hkey=22da1f28-f457-45d4-8f13-583685cd7b6a.

[43] European Workshop on Structural Health Monitoring. (2018). https://www.bindt.org/events/PastEvents/ewshm-2018/.

[44] 13th International Conference on Applications of Statistics and Probability in Civil Engineering. (2019). https://www.icasp13.snu.ac.kr/.

[45] Ostachowicz, W. and Güemes, A. (eds.). (2013). New Trends in Structural Health Monitoring (Vol. 542). Springer Science & Business Media.

[46] Chen, H. P. (2018). Structural Health Monitoring of Large Civil Engineering Structures. John Wiley & Sons.

Chapter 2

Mathematical Preliminaries

This chapter presents an overall introduction to some of the core ideas which form the building blocks of the real-time detection approach presented in this book. Some of these buildings blocks are quite well-known, while some others are more specific. For the purpose of an overall completeness, we have introduced them briefly without the intention to be comprehensive. References are indicated for the readers to engage with these topics in detail.

2.1 Deterministic and random signals

Most signals dynamics and vibrations are inherently connected to a physical phenomenon or, a physical system. These signals can be measured using sensors and are often referred to as observed data or, time historics. Examplcs of signals include temperature fluctuations in a room as a function of time, voltage variations from a vibration transducer, pressure changes at a point in an acoustic field, displacements and accelerations of a vibrating body.

Time histories can be broadly categorized as deterministic or random. Deterministic signals are those whose behaviour can be predicted exactly. For example, the response of a single mass–spring oscillator (often idealized as a single degree of freedom system) undergoing free vibration is deterministic. Deterministic signals arc furthcr catcgorizcd as pcriodic and non pcriodic as shown in Fig. 2.1.

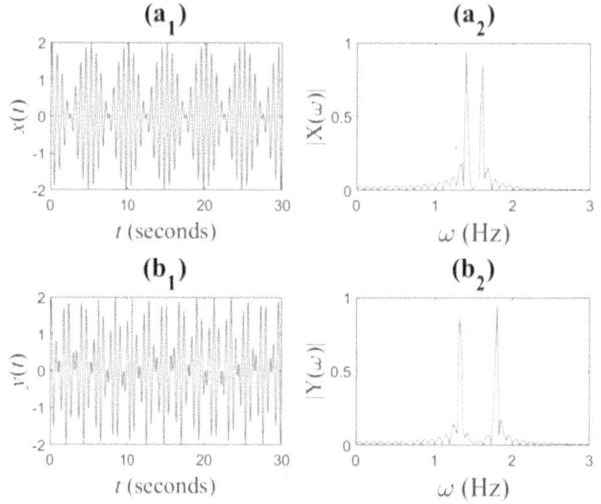

Figure 2.1: Periodic and almost periodic (aperiodic) signals. The subfigures a_1 and a_2 represent a periodic signal and its Fourier transform (The frequencies are harmonically related as they are integral multiples of 0.1 Hz) The subfigures b_1 and b_2 represent aperiodic signals since their frequencies are not integral multiples of 0.1 Hz.

Often, signals are not deterministic, which means that the value of a signal is not exactly known given a certain time instant in advance. Under such circumstances, one cannot write explicit time descriptions for them. These are often referred to as random signals [5]. Random signals are often described in the context of probability theory as stochastic processes. Loosely speaking, a stochastic process is described using an ensemble, i.e., random variables at different instants of time. Consequently, they are conveniently expressed using first and second order probability density functions $p(x(t_1);t_1)$ and $p(x(t_1),x(t_2);t_1,t_2)$ respectively. The first and second order statistics can then be expressed as [5]:

$$\mu_x(t) = E\{x(t)\} = \int_{-\infty}^{\infty} x(t)p(x;t)dx$$

$$R_{xx}(t_1,t_2) = \int_{-\infty}^{\infty}\int_{-\infty}^{\infty} x(t_1)x(t_2)p(x(t_1),x(t_2);t_1,t_2)dx_1dx_2 \tag{2.1}$$

In some signals, the statistical properties, namely, the mean and the autocorrelation $R_{xx}(t_1,t_2)$, do not vary over time and are referred to as weakly stationary signals. Furthermore, since measuring the ensembles are impossible in practical systems, ergodicity assumptions are often used to replace the expectations with sample statistics. Thus the temporal statistics can be expressed as [3]:

$$\bar{y} = \lim_{T\to\infty} \frac{1}{T} \int_0^T y(t)dt$$

$$\bar{y^2} = \lim_{T\to\infty} \frac{1}{T} \int_0^T y^2(t)dt$$

$$y_{rms} = \sqrt{\bar{y^2}} = \sqrt{\lim_{T\to\infty} \frac{1}{T} \int_0^T y^2(t)dt} \tag{2.2}$$

Of utmost importance in random signal processing is a special type of process called white noise process or simply white noise [3]. White noise can be defined as a random process that is characterized by constant power at all frequencies (ref PSDF) or, a Dirac delta $\delta(\tau)$ autocorrelation function. This is clearly shown in Fig. 2.2.

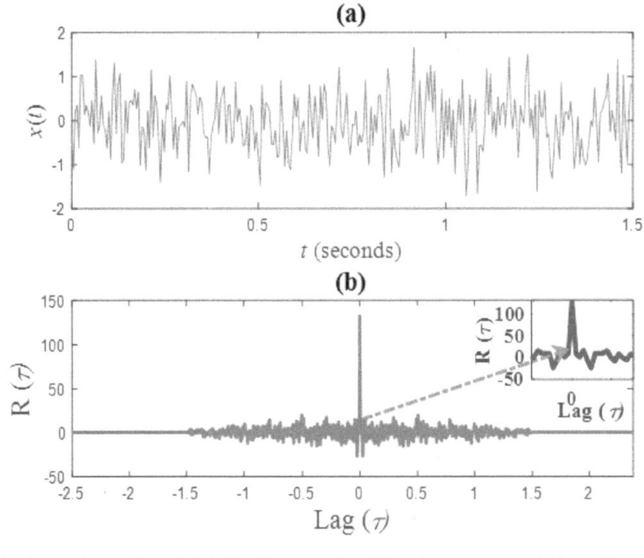

Figure 2.2: A random signal (a) is characterized by its autocorrelation function R(τ) (b) represented by Dirac delta function $\delta(\tau)$ at ($\tau = 0$) (in a limiting process).

2.2 Fourier transform

Any periodic function $f(x)$ with time period T can be expressed as an infinite trigonometric series of the form [1]:

$$X(t) = \frac{a_0}{2} + \sum_{n=1}^{\infty} \left[a_n cos\left(\frac{2\pi nT}{T}\right) + b_n sin\left(\frac{2\pi nT}{T}\right) \right] \tag{2.3}$$

where, a_0, a_n and b_n are the coefficients of the Fourier series. This series can also be expressed in complex notation as:

$$X(t) = \sum_{n=-\infty}^{\infty} C_n \mathrm{e}^{\frac{i2\pi n}{T}} \Longrightarrow C_n = \frac{1}{T} \sum_{n=-\infty}^{\infty} X(t) \mathrm{e}^{-\frac{i2\pi n}{T}} \tag{2.4}$$

where, C_n is the coefficient of Fourier series. Defining $\omega_n = \frac{2\pi n}{T}$ as the frequency of the kth harmonic and $\Delta\omega_k = \frac{2\pi}{T}$ to be the spacing between adjacent harmonics, it can be observed that for a large time period $T \to \infty$, the spacing between frequencies will become inseparable ($\Delta\omega \to 0$). In this case $X(t)$ will be no longer be periodic and the Fourier series turns into a Fourier integral and the Fourier coefficients turn into continuous functions of frequency called Fourier transforms [1].

$$X(\omega) = \int_{-\infty}^{\infty} X(t) \mathrm{e}^{-i\omega t} dt \tag{2.5}$$

Analogous to the Fourier series an inverse Fourier transform $X(t)$ can be defined as:

$$X(t) = \int_{-\infty}^{\infty} X(\omega) \mathrm{e}^{i\omega t} d\omega \tag{2.6}$$

Together Eqs. 2.7 and 2.6 are referred to as Fourier transform pairs. Numerically a Fourier transform is evaluated by discretizing the time interal using sampled data ($\Delta, 2\Delta, 3\Delta, \ldots$) and also at frequencies $k/N\Delta$ resulting in a $N-$point Fourier transform, also called discrete Fourier transform or, DFT.

For a 5 DOF linear system, a typical FFT plot is expected to indicate 5 peaks corresponding to the energy content of the spectrum. A simple example is provided below (Fig. 2.3).

Parseval's identity

Parseval's theorem in the context of Fourier transform is an important result that loosely states that the average power of the signal $x(t)$ is equal to the sum of the power associated with individual frequency components. Mathematically, this statement translates to [5]:

$$\frac{1}{T} \int_{0}^{T} X(t)^2 dt = \left| C_n^2 \right| \tag{2.7}$$

Parseval's identity however relates to the equality of the average power in the time-domain and the corresponding counterpart in the frequency domain.

Nyqist samping theorem

Since the evaluation of DFT involves sampling the time interval and evaluation of the Fourier integral both in discrete instances of time as well as frequency, it

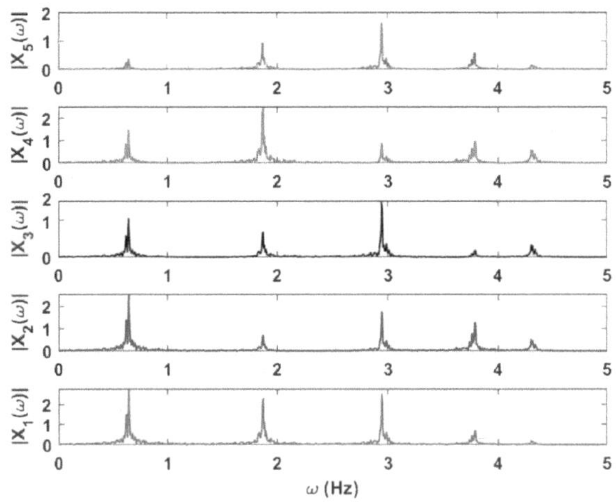

Figure 2.3: A typical FFT plot for a 5 DOF linear system.

is important to abide by certain restrictions on sampling frequency $f_s = 1/\Delta$. The Nyquist-Shannon sampling theorem loosely states that if a function $x(t)$ contains frequencies lesser than, or, equal to F hertz, then it can be completely sampled by a series of points spaced $\Delta = 1/2F$ seconds apart. A sufficient sample-rate is therefore anything larger than $2F$ samples per second, and on other hand a sampling rate less than $2F$ samples per second leads to serious issues called aliasing.

Uncertainty principle
This principle pertains to the fundamental limitation of Fourier transform, that, in order to obtain an arbitrarily finely resolved Fourier transform, increasing the localization in the time domain will not result in a superior resolution in the frequency domain, and vice versa. In other words, one cannot achieve an arbitrarily fine resolution in both the time and frequency domains simultaneously. More formally, it can be stated as [5]:

$$\Delta\omega\Delta t \geq \frac{1}{2} \tag{2.8}$$

The Eq. 2.8 is also called the bandwidth-time (BT) product. For example, for the rectangular window the BT product is 2π nd the Gaussian pulse e^{-at^2} has a minimum BT product of $1/2$.

Windowing
Windowing of a signal can impact the computed estimates of FFT and is a function of the type of window and how the computation of Fourier Transform is

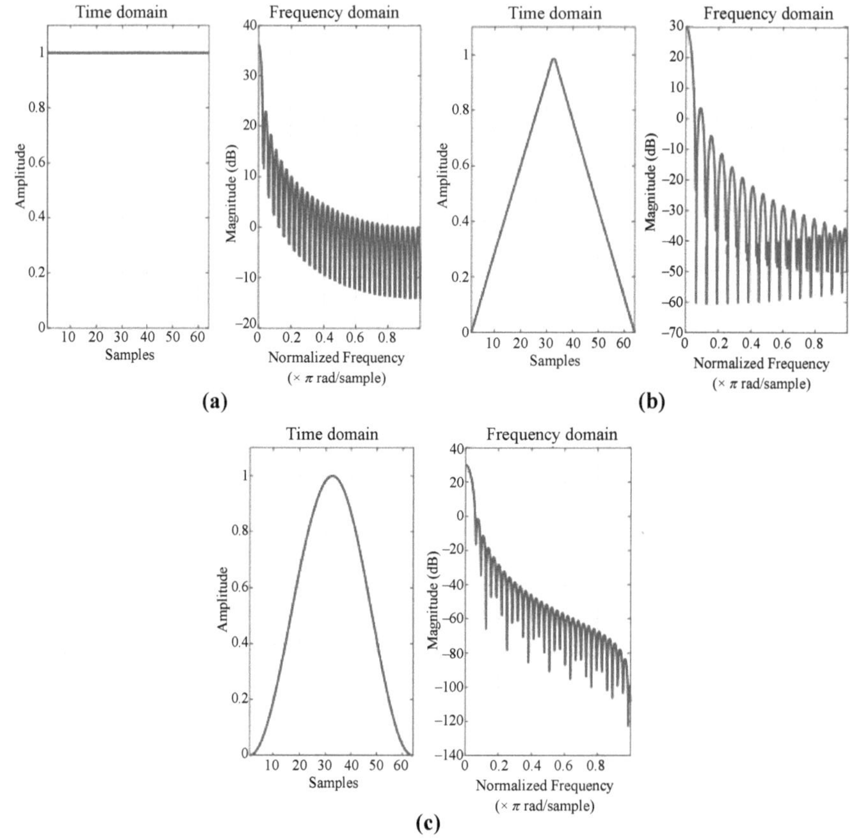

Figure 2.4: Representation of (a) Rectangular Window, (b) Bartlett Window (c) Hanning Window in Time and Frequency Domain, respectively.

implemented. A detailed example can be obtained in [13]. The signal $y(t)$ can be multiplied with a window $w(t)$ to obtained a windowed $y_w(t) = y(t)w(t)$. As two disparate examples, Fig. 2.4(a) and 2.4(b) show the time and frequency domain representation of Bartlett $(W(t) = 1 - |t|N \; for \; |t| \leq N \quad W(t) = 0, \; for \; |t| > N)$ and Fig. 2.4(c) represents Hanning window $(W(t) = \cos^2(t/N))$, respectively.

The frequency domain response has been plotted against a frequency range from 0 to a normalized value of 1. This range can be appropriately changed to a desired range by multiplying with a scalar conforming to the sampling incorporated in the time domain. The leakage factor is defined as the ratio of the power of the side-lobes to the total window power. The Bartlett window is known to have a leakage factor larger than the smooth windows and is traditionally undesirable for many applications. The relative side-lobe attenuation, i.e., the difference of height between the main-lobe peak and the highest side-lobe peak is also greater for smoother windows.

2.3 Frequency response function

Consider the equation of motion of a SDOF system with mass **M**, stiffness **K** and damping **C** as:

$$\mathbf{M}\mathbf{X}(\mathbf{t}) + \mathbf{C}\mathbf{X}(\mathbf{t}) + \mathbf{K}\mathbf{X}(\mathbf{t}) = \mathbf{F}(\mathbf{t}) \tag{2.9}$$

For the linear system in Eq. (6.1) subjected to unit amplitude harmonic input, $F(t) = \mathrm{e}^{i\omega t}$, the corresponding output $X(t)$ is given as [2]:

$$X(t) = \mathbf{H}(\omega)\mathrm{e}^{\mathbf{i}\omega\mathbf{t}} \tag{2.10}$$

where, $\mathbf{H}(\omega)$ is the frequency response function of the system evaluated at angular frequency ω. Taking the Fourier transform of Eq. (6.1) and setting the initial conditions to zero (0), the expression for $H(\omega)$ can be derived as:

$$\begin{aligned} -\mathbf{M}\omega^2\mathbf{X}(\omega) + \mathbf{i}\mathbf{C}\omega\mathbf{X}(\omega) + \mathbf{K}\mathbf{X}(\omega) &= \mathbf{F}(\omega) \\ \mathbf{H}(\omega) = \frac{\mathbf{X}(\omega)}{\mathbf{F}(\omega)} &= \frac{\mathbf{1}}{-\mathbf{M}\omega^2 + \mathbf{i}\mathbf{C}\omega + \mathbf{K}} \end{aligned} \tag{2.11}$$

The frequency response function, $bfH(\omega)$ can be utilized for the computation of the power spectral density function (PSDF) and the mean square response of any randomly excited dynamical system.

2.4 Impulse response function

An impulsive force is a large force that acts for a very short duration of time. The impulse force can be modelled using the Dirac delta function which is defined as follows,

$$\delta(t-\tau) = \begin{cases} +\infty; & t = \tau \\ 0; & t \neq \tau \end{cases} \tag{2.12}$$

The delta function has the important property, known as the magnitude of the impulse. For a unit impulse, the area under the curve is given as,

$$\int_{-\infty}^{\infty} \delta(t-\tau)\, dt = 1 \tag{2.13}$$

The response to a unit impulse load is defined as the impulse response function which is generally denoted by $h(t)$. Consider the input force in Eq. (6.1) as the impulse force $\delta(t)$, i.e., $F(t) = \delta(t)$. The impulse response function $h(t)$ is the solution of the following equation:

$$\mathbf{M}h(t) + \mathbf{C}h(t) + \mathbf{K}h(t) = \delta(t) \tag{2.14}$$

Taking the Fourier transform of the impulse response function $h(t)$, the frequency response function $\mathbf{H}(\omega)$ can be obtained as [2]:

$$\mathbf{H}(\omega) = \int_{-\infty}^{\infty} \mathbf{h(t)} \exp(-\mathbf{i}\omega\mathbf{t})\mathbf{dt} \tag{2.15}$$

Thus it is clear from Eq. 2.15, that the frequency and impulse response functions are nothing but Fourier transform pairs.

2.5 Power spectral density

Power spectral density function (**PSDF**) $S_x(\omega)$ denotes the distribution of energy of a signal over a range of frequency. It is estimated by taking the Fourier transform of the auto-correlation function (R_X) and can be expressed as [3]:

$$S_X(\omega) = \int_{-\infty}^{\infty} R_X(\tau)\mathrm{e}^{-i\omega t} d\tau \tag{2.16}$$

where the autocorrelation function of a wide sense stationary process $X(t)$ can be defined as [3]: $R_X(\tau) = E\{X(t)X(t+\tau)\}$ The autocorrelation of a white noise process (with variance σ^2) is $\sigma^2\delta(\tau)$ where, $\delta(\tau)$ Dirac delta function and the PSDF of $\sigma^2\delta(\tau)$ yields σ^2. This underscores an important result that is the PSDF of a white noise process is a constant value at all the frequencies, thereby, signifying constant power at all the frequencies. A typical PSDF plot for a 5 DOF linear system is illustrated in Fig. 2.5.

Consider an input signal with PSDF defined by $S_F(\omega)$. The PSDF of the output $S_X(\omega)$ of a system can be estimated with the help of frequency response function $H(\omega)$ using the following relationship [2]:

$$S_X(\omega) = H^*(\omega)H(\omega)S_X(\omega) = |H(\omega)|^2 S_X(\omega) \tag{2.17}$$

2.6 Eigenperturbation–a bird's eye view

Perturbation can be defined as a deviation from the normal functioning of a system, usually influenced by an external entity. Research related to eigenperturbation popularized by Lord Rayleigh [4] can be structured as obtaining eigenvalues and eigenvectors of a system *perturbed* from one known set of eigenvalues

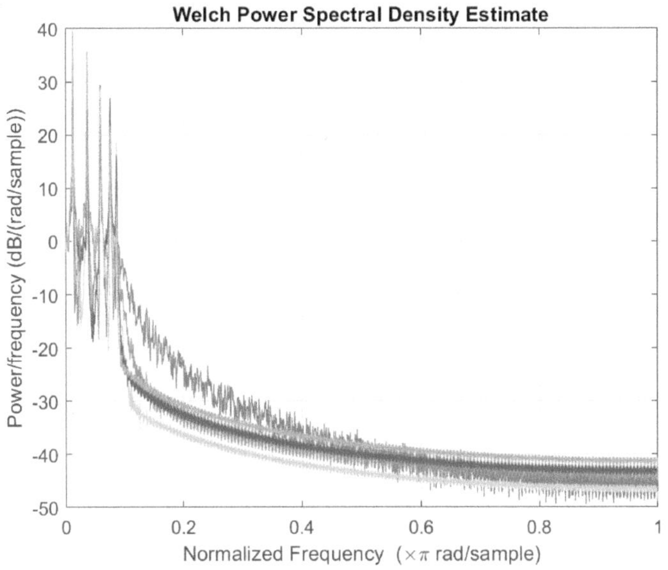

Figure 2.5: A PSDF plot for a 5 DOF linear system.

and eigenvectors (collectively known as the *eigenspace*). Based on this concept, many researchers have provided their own versions of eigenperturbation – broadening the domain from projector operator formalism by Löwdin [6] to providing realistic perturbation bounds as norms [7]. Existing literature reveals that fundamentals of eigenperturbation can be realized from an overarching framework that explicitly deals with higher order deviations. In this book, the treatment of eigenperturbation will be demonstrated from a low-rank eigenspace update – essentially implying that first order eigenperturbation (FOEP) will be employed for all investigations (both numerical and experimental). However, it becomes important to appreciate the fact that FOEP can be specifically obtained from a wide genre of higher order perturbation – which is essentially the key objective of this section.

Recent development of eigenperturbation based real-time health monitoring algorithms, such as RPCA [6], RSSA [4] and RCCA [7] are mainly centred around recursive estimation of the data covariance matrix and its eigen decomposition in real-time. This provides an alternative to avoiding the update of the full-rank covariance matrix and its eigen-decomposition simultaneously, thereby reducing the computational complexity. Prior to going into the details of the recursive algorithms, it is important to discuss generalized higher order perturbation theory and how FOEP can be derived as a special case.

2.7 Recursive covariance estimation in a higher order perturbation framework

A standard form of the eigen equation can be derived based on the dynamical behaviour of a system. The discussed approach being an output-only technique, inverse vibration formulations lead to the standard eigen equation as:

$$\mathbf{C}_0\mathbf{V}_0 = \mathbf{\Lambda}_0\mathbf{V}_0 \tag{2.18}$$

where, $\mathbf{C_0}$ is the covariance matrix composed of the physical responses streamed at the start of monitoring, $\mathbf{V_0}$ and $\mathbf{\Lambda_0}$ are the eigen vector and eigen value matrices – collectively termed as the *eigenspace* [6, 9, 10].

Consider the rank-n update of the eigenspace that provides the higher eigen-perturbation terms. The following definitions for the nth order update of the matrices involved in Eq. 2.18 will be useful for subsequent derivations:

$$\left.\begin{array}{l}\mathbf{C}_n = \mathbf{C}_0 + \delta\mathbf{C} + \delta^2\mathbf{C} + \delta^3\mathbf{C} + \ldots + \delta^n\mathbf{C} \\ \mathbf{V}_n = \mathbf{V}_0 + \delta\mathbf{V} + \delta^2\mathbf{V} + \delta^3\mathbf{V} + \ldots + \delta^n\mathbf{V} \\ \mathbf{\Lambda}_n = \mathbf{\Lambda}_0 + \delta\mathbf{\Lambda} + \delta^2\mathbf{\Lambda} + \delta^3\mathbf{\Lambda} + \ldots + \delta^n\mathbf{\Lambda}\end{array}\right\} \tag{2.19}$$

Substituting the above values in Eq. 2.18, the nth order update equations can be therefore, formulated as:

$$\begin{array}{l}\left(\mathbf{C}_0 + \delta\mathbf{C} + \delta^2\mathbf{C} + \delta^3\mathbf{C} + \ldots + \delta^n\mathbf{C}\right) = \left(\mathbf{V}_0 + \delta\mathbf{V} + \delta^2\mathbf{V} + \delta^3\mathbf{V} + \ldots + \delta^n\mathbf{V}\right) \\ \left(\mathbf{\Lambda}_0 + \delta\mathbf{\Lambda} + \delta^2\mathbf{\Lambda} + \delta^3\mathbf{\Lambda} + \ldots + \delta^n\mathbf{\Lambda}\right)\left(\mathbf{V}_0 + \delta\mathbf{V} + \delta^2\mathbf{V} + \delta^3\mathbf{V} + \ldots + \delta^n\mathbf{V}\right)^T\end{array} \tag{2.20}$$

A detailed expansion of the terms is not provided for brevity. However, the comparison of similar order terms reveals the following equation:

$$\begin{array}{l}\delta^n\mathbf{C} = \mathbf{V}_0\mathbf{\Lambda}_0\delta\mathbf{V}^T + \mathbf{V}_0\left[\sum\limits_{i=1}^{k}(\delta\mathbf{\Lambda})^i\left(\delta\mathbf{V}^T\right)^{n-i}\right] + \mathbf{V}_0\delta^n\mathbf{\Lambda}\mathbf{V}_0^T + \delta\mathbf{V}\mathbf{\Lambda}_0\delta^{n-1}\mathbf{V}^T + \\ \delta\mathbf{V}\left[\sum\limits_{i=1}^{k}(\delta\mathbf{\Lambda})^i\left(\delta\mathbf{V}^T\right)^{n-(i+1)}\right] + \delta^2\mathbf{V}\mathbf{\Lambda}_0\delta^{n-2}\mathbf{V}^T + \delta^2\mathbf{V}\left[\sum\limits_{i=1}^{k}(\delta\mathbf{\Lambda})^i\left(\delta\mathbf{V}^T\right)^{n-(i+2)}\right] + \\ \delta^3\mathbf{V}\mathbf{\Lambda}_0\delta^{n-3}\mathbf{V}^T + \delta^3\mathbf{V}\left[\sum\limits_{i=1}^{k}(\delta\mathbf{\Lambda})^i\left(\delta\mathbf{V}^T\right)^{n-(i+3)}\right] + \delta^n\mathbf{V}\mathbf{\Lambda}_0\mathbf{V}_0^T\end{array} \tag{2.21}$$

A simple expression in terms of discrete summation can be obtained from the above, as follows:

$$\begin{aligned}
&\delta^n\mathbf{C} = \mathbf{V}_0\boldsymbol{\Lambda}_0\delta\mathbf{V}^T + \mathbf{V}_0\delta\boldsymbol{\Lambda}\delta^{n-1}\mathbf{V}^T + \mathbf{V}_0\delta^2\boldsymbol{\Lambda}\delta^{n-2}\mathbf{V}^T + \mathbf{V}_0\delta^3\boldsymbol{\Lambda}\delta^{n-3}\mathbf{V}^T + \\
&\ldots + \mathbf{V}_0\delta^k\boldsymbol{\Lambda}\delta^{n-k}\mathbf{V}^T + \mathbf{V}_0\delta^n\boldsymbol{\Lambda}\mathbf{V}_0^T + \delta\mathbf{V}\boldsymbol{\Lambda}_0\delta^{n-1}\mathbf{V}^T + \delta\mathbf{V}\delta\boldsymbol{\Lambda}\delta^{n-2}\mathbf{V}^T + \\
&\delta\mathbf{V}\delta^2\boldsymbol{\Lambda}\delta^{n-3}\mathbf{V}^T + \delta\mathbf{V}\delta^3\boldsymbol{\Lambda}\delta^{n-4}\mathbf{V}^T + \ldots + \delta\mathbf{V}\delta^k\boldsymbol{\Lambda}\delta^{n-(k+1)}\mathbf{V}^T + \delta^2\mathbf{V}\boldsymbol{\Lambda}_0\delta^{n-2}\mathbf{V}^T + \\
&\delta^2\mathbf{V}\delta\boldsymbol{\Lambda}\delta^{n-3}\mathbf{V}^T + \delta^2\mathbf{V}\delta^2\boldsymbol{\Lambda}\delta^{n-4}\mathbf{V}^T + \delta^2\mathbf{V}\delta^3\boldsymbol{\Lambda}\delta^{n-5}\mathbf{V}^T + \ldots + \delta^2\mathbf{V}\delta^k\boldsymbol{\Lambda}\delta^{n-(k+2)}\mathbf{V}^T \\
&+ \delta^3\mathbf{V}\boldsymbol{\Lambda}_0\delta^{n-3}\mathbf{V}^T + \delta^3\mathbf{V}\delta\boldsymbol{\Lambda}\delta^{n-4}\mathbf{V}^T + \delta^3\mathbf{V}\delta^2\boldsymbol{\Lambda}\delta^{n-5}\mathbf{V}^T + \delta^3\mathbf{V}\delta^3\boldsymbol{\Lambda}\delta^{n-6}\mathbf{V}^T + \ldots + \\
&\delta^3\mathbf{V}\delta^k\boldsymbol{\Lambda}\delta^{n-(k+3)}\mathbf{V}^T + \delta^n\mathbf{V}\boldsymbol{\Lambda}_0\mathbf{V}_0^T \\
&= \mathbf{V}_0\left[\sum_{i=1}^{k}(\delta\boldsymbol{\Lambda})^i\left(\delta\mathbf{V}^T\right)^{n-i}\right] + \mathbf{V}_0\delta^n\boldsymbol{\Lambda}\mathbf{V}_0^T + \delta\mathbf{V}\boldsymbol{\Lambda}_0\delta^{n-1}\mathbf{V}^T + \\
&\delta\mathbf{V}\left[\sum_{i=1}^{k}(\delta\boldsymbol{\Lambda})^i\left(\delta\mathbf{V}^T\right)^{n-(i+1)}\right] + \delta^2\mathbf{V}\boldsymbol{\Lambda}_0\delta^{n-2}\mathbf{V}^T + \delta^2\mathbf{V}\left[\sum_{i=1}^{k}(\delta\boldsymbol{\Lambda})^i\left(\delta\mathbf{V}^T\right)^{n-(i+2)}\right] + \\
&\delta^3\mathbf{V}\boldsymbol{\Lambda}_0\delta^{n-3}\mathbf{V}^T + \delta^3\mathbf{V}\left[\sum_{i=1}^{k}(\delta\boldsymbol{\Lambda})^i\left(\delta\mathbf{V}^T\right)^{n-(i+3)}\right] + \delta^n\mathbf{V}\boldsymbol{\Lambda}_0\mathbf{V}_0^T
\end{aligned} \tag{2.22}$$

The expressions for first order eigenperturbation (FOEP) can be *intuitively* obtained by substituting n=1 in Eq. 2.22 [10]:

$$\delta\mathbf{C} = \mathbf{V}_0\boldsymbol{\Lambda}_0\delta\mathbf{V}^T + \mathbf{V}_0\delta\boldsymbol{\Lambda}\mathbf{V}_0^T + \delta\mathbf{V}\boldsymbol{\Lambda}_0\mathbf{V}_0^T + \mathcal{O}\left(\delta^n\right) \tag{2.23}$$

For FOEP, perturbation order greater than *one* does not exist (which automatically cancels δ^n terms ($\forall n > 1$), [8]). The cross-matrix multiplication terms are also eliminated in this process - however, spectrally decomposable terms remain intact. An investigation into the perturbation expansion reveals that the expressions obtained above are quite accurate, considering the low magnitude of the perturbation matrices [7]. Distinct eigenvalues initiate the separation of the spectral terms from its neighbors. If this separation ($m-n$ columns of $\mathbf{C}$) is denoted by Δ, then the second order error term is bounded by $||\delta\mathbf{C}||^2/\boldsymbol{\Delta}$. In engineering practice $||\delta\mathbf{C}||/\boldsymbol{\Delta}$ should be considerably less than one – say one-hundredth and beyond – before one starts to trust this approximation [8]. This result should be treated as a caution for easy rank detection through small singular values. In the presence of noise, singular values tend to proportionately amplify under higher order perturbations (order of $\sqrt{n}$). In practice, this corresponds to acquiring an incorrect rank estimation of the matrices, especially when the signal sound-to-noise ratio approaches $\sqrt{n}$.

2.8 A note on eigenvalue sensitivity

An important framework for eigenvalue computation is to produce a sequence of similarity transformations for enforcing strictly diagonally dominant (SDD) matrices. The configuration of these matrices requires the investigation of individual elements towards approximating the eigenvalues - i.e., how well do the diagonal elements of the matrix resemble the eigenvalues. This requires consideration of the error norms based on established perturbation theorems which are explained as follows. Throughout this article the notation $||.||$ will denote a vector norm when applied to a vector and an operator matrix norm when applied to a matrix.

Theorem 1: *Gershgorin's Circle Theorem.*
Let $\mathbf{A}$ be an arbitrary $n \times n$ matrix with λ as any one of the eigenvalues of $\mathbf{A}+\mathbf{E}$, where $\mathbf{E}$ is a small perturbation with zero diagonal entries. The spectral decomposition of the matrix can be expressed as:

$$\mathbf{A}+\mathbf{E} = \mathbf{V}\mathbf{\Lambda}\mathbf{V}^{-1} \tag{2.24}$$

The mathematical identity using Gershgorin theorem provides the equation:

$$\lambda(\mathbf{\Lambda}) \subseteq \bigcup_{i=1}^{n} \mathbf{A}_i \tag{2.25}$$

where, $\mathbf{A}_i = \left\{ z \in \Re : |z - a_i| \leq \sum_{j=1}^{n} |e_{ij}| \right\}$

Proof:

If $\mathbf{F} \in \Re^{n \times n}$ and $||\mathbf{F}||_p < 1$, then $\mathbf{I}-\mathbf{F}$ is nonsingular and

$$(\mathbf{I}-\mathbf{F})^{-1} = \sum_{k=0}^{\infty} \mathbf{F}^k \tag{2.26}$$

with,

$$||(\mathbf{I}-\mathbf{F})^{-1}||_p \leq \frac{1}{1-||\mathbf{F}||_p} \tag{2.27}$$

The proof starts with considering a contradiction that $(\mathbf{I}-\mathbf{F})$ is singular - which establishes that $(\mathbf{I}-\mathbf{F})x = 0$ for any nonzero x. This means that $||x||_p = ||\mathbf{F}x||_p$ suggesting that $||\mathbf{F}||_p \geq 1$ – is a contradiction. Therefore, $\mathbf{I}-\mathbf{F}$ is nonsingular. The inverse of the matrix can be obtained by considering the identity:

$$\left(\sum_{k=0}^{N} \mathbf{F}^k \right) (\mathbf{I}-\mathbf{F}) = \mathbf{I} - \mathbf{F}^{N+1} \tag{2.28}$$

Since $||\mathbf{F}||_p < 1$, it follows that $\lim_{k \to \infty} \mathbf{F}^k = 0$ as $||\mathbf{F}_k||_p \leq ||\mathbf{F}||_p^k$. Therefore, the following equation holds.

$$\left(\lim_{N \to \infty} \sum_{k=0}^{N} \mathbf{F}^k \right) (\mathbf{I}-\mathbf{F}) = \mathbf{I} \tag{2.29}$$

It follows that $(\mathbf{I}-\mathbf{F})^{-1} = \lim_{N\to\infty} \sum_{k=0}^{N} \mathbf{F}^k$. The proof of this lemma can be concluded by easily manifesting the equation in the form:

$$\left\|(\mathbf{I}-\mathbf{F})^{-1}\right\|_p \le \sum_{k=0}^{\infty} \|\mathbf{F}\|_p^k = \frac{1}{1-\|\mathbf{F}\|_p} \tag{2.30}$$

Without loss of generality, suppose $\lambda \in \lambda(\mathbf{\Lambda})$ for which $\lambda \neq a_i\ \forall i = 1:n$. As $(\mathbf{A}-\lambda\mathbf{I})+\mathbf{E}$ is singular, it follows from Lemma 1 (**Appendix A**) that:

$$1 \le \left\|(\mathbf{A}-\lambda\mathbf{I})^{-1}\mathbf{E}\right\|_\infty = \sum_{j=1}^{n} \frac{|e_{kj}|}{|a_k-\lambda|} \tag{2.31}$$

$\forall k = 1:n$. However, this implies that $\lambda \in \mathbf{A}_k$. The proof establishes that the diagonal portion of the covariance matrix so obtained approximates the eigenvalue of the matrix. It follows that the *matrix bears an SDD structure*, which enables recursive updates of the eigenspace at each instant. This forms the key cornerstone of FOEP analyses that enables online recovery of structural modes from a dynamic system.

Theorem 2: *Bauer-Fike Theorem.*
Although the Bauer-Fike theorem discusses the eigenvalue perturbations of diagonalizable matrices (which are mostly SDD), it essentially studies the localization of eigenvalues that are concentrated within small regions of the complex plane. This theorem can be considered as a generalization of the Gershgorin's circle theorem – but for the purposes of this article, discussions are limited to operator norms, derived from related vector norms.

Theorem: Let $\mathbf{A}$ be an arbitrary $n \times n$ diagonalizable matrix satisfying $\mathbf{A} = \mathbf{V\Lambda V}^{-1}$ and let $\mathbf{E}$ be a small perturbation matrix. Every eigenvalue μ of the matrix $\mathbf{A}+\mathbf{E}$ satisfies:

$$\min_{\lambda\in\lambda(A)} |\mu-\lambda| \le \|\mathbf{V}\| \cdot \left\|\mathbf{V}^{-1}\right\| \cdot \|\mathbf{E}\| \tag{2.32}$$

where λ is some eigenvalue of $\mathbf{A}$.

Proof:
A straightforward explanation of the theorem would suggest its veracity if $\lambda \in \lambda(\mathbf{\Lambda})$. Contrary to this idea, considering the singularity of $\mathbf{V}^{-1}(\mathbf{A}+\mathbf{E}-\lambda\mathbf{I})\mathbf{V}$, $\mathbf{I}+(\mathbf{A}-\lambda\mathbf{I})^{-1}(\mathbf{V}^{-1}\mathbf{EV})$ also becomes singular. From Lemma 1, the following inequality is obtained:

$$1 \le \left\|(\mathbf{A}-\lambda\mathbf{I})^{-1}\left(\mathbf{V}^{-1}\mathbf{EV}\right)\right\|_p \le \left\|(\mathbf{A}-\lambda\mathbf{I})^{-1}\right\|_p \|\mathbf{X}\|_p \|\mathbf{E}\|_p \left\|\mathbf{X}^{-1}\right\|_p \tag{2.33}$$

As $(\mathbf{A}-\lambda\mathbf{I})^{-1}$ is diagonal and the p-norm of a diagonal matrix automatically translates to the absolute value of the largest diagonal entry, and it follows that

$$\left\|(\mathbf{A}-\lambda\mathbf{I})^{-1}\right\|_p = \min_{\lambda \in \lambda(A)} \frac{1}{|\mu - \lambda|} \tag{2.34}$$

which completes the proof. If $||\mathbf{E}||$ is small, then the theorem states that the eigenvalues of $\mathbf{A}+\mathbf{E}$ are close to that of $\mathbf{A}$. Interestingly, the distance between the eigenvalues of $\mathbf{A}+\mathbf{E}$ and $\mathbf{A}$ vary linearly with the perturbation matrix $\mathbf{E}$, which forms yet another fundamental hypothesis of eigenperturbation theory.

Certain cases require the perturbation bounds to be computed on singular vectors. Since the individual singular vectors corresponding to a cluster of singular values is unstable, bounds for the subspace spanned by the singular vectors – known as the *singular subspace* – are computed. For two one dimensional subspaces $\mathcal{X}$ and $\mathcal{Y}$, the angle between the subspaces is given by: $\theta = \cos^{-1}\left|x^T y\right|$, where x and y are the vectors of norm one spanning $\mathcal{X}$ and $\mathcal{Y}$. In a more generalized sense, for complex dynamical systems, this construction can be expanded to subspaces of dimension k. Considering x and Y to be the orthonormal bases for $\mathcal{X}$ and $\mathcal{Y}$ respectively, the above expression can be formulated as:

$$\theta_i|_k = \cos^{-1}|\gamma_i|_k \tag{2.35}$$

where, $\gamma_1 \geq \gamma_2 \geq \ldots \geq \gamma_i$ are the singular values of $X^T Y$. The formulated measure is then referred to as *canonical angles* that provide a distance metric between the subspaces. An important step in identifying the modal structure from output responses involves projecting a source p to a 1D orthogonal space (pre-whitening step) where the pth component resides. It is therefore, worthwhile to note the connection between canonical angles and projections. Consider P_X and P_Y to be the orthogonal projections onto $\mathcal{X}$ and $\mathcal{Y}$. If $\mathcal{X}$ and $\mathcal{Y}$ are equal, it naturally follows that $P_X = P_Y$, which retains $||P_X = P_Y||$ as the distance metric. However, for a generalized treatment of the random subspaces, it can be shown that the Frobenius norm is related to the canonical projection as:

$$||P_X = P_Y||_F = \sqrt{2}||\sin\theta||_F \tag{2.36}$$

Thus the two measures decay to zero at the same rate. This allows for the systematic manifestation of orthonormal projections that form the preliminary step of the recursion algorithm.

2.9 Gershgorin's theorem

One of the primary objectives of a preliminary dynamic analysis of a vibrating system is to evaluate the natural frequencies of each of its components. The trace of a matrix merely provides the sum of its eigenvalues and does not provide insights about the range for those eigenvalues 2.6. One of the easier methods is to employ Gershgorin's theorem for the same, which is subsequently discussed in the purview of the dynamics of a structural system.

Prior to getting into the details of the theorem, it is important to review certain key concepts that are essential for its theoretical treatment. The condition in which the individual elements of a matrix $\mathbf{A}_{nn}$ follow the relation

$$\mathbf{A}_{nn}$$
$$|\mathbf{A}_{ii}| > \sum_{j \neq i} |\mathbf{A}_{ij}|, \; for\; i = 1, 2, ..., n \tag{2.37}$$

render the matrix as a *strictly diagonally dominant* (SDD) matrix. On employing the non-singularity of such type of matrices, it now becomes easy to understand the Gershgorin's theorem.

The theorem states that, every eigenvalue of the SDD matrix $\mathbf{A}_{nn}$ satisfies:

$$|\lambda - \mathbf{A}_{ii}| > \sum_{j \neq i} |\mathbf{A}_{ij}|, \; i \in \{1, 2, ..., n\} \tag{2.38}$$

The proof of this theorem can be stated as follows:

Suppose that λ is an eigenvalue of the matrix $\mathbf{A}$. Then the matrix $\lambda\mathbf{I} - \mathbf{A}$ is SDD if $|\lambda - \mathbf{A}_{ii}| > \sum_{j \neq i} |\mathbf{A}_{ij}|$ for every i. If Eq. 2.38 is not satisfied then $\lambda\mathbf{I} - \mathbf{A}$ is SDD. If $\lambda\mathbf{I} - \mathbf{A}$ is SDD, it automatically implies that the matrix is nonsingular and as a result, λ is not an eigenvalue. For λ to be an eigenvalue, the above stated rule must hold, which forms the key basis of the Gershgorin's theorem. In analyzing

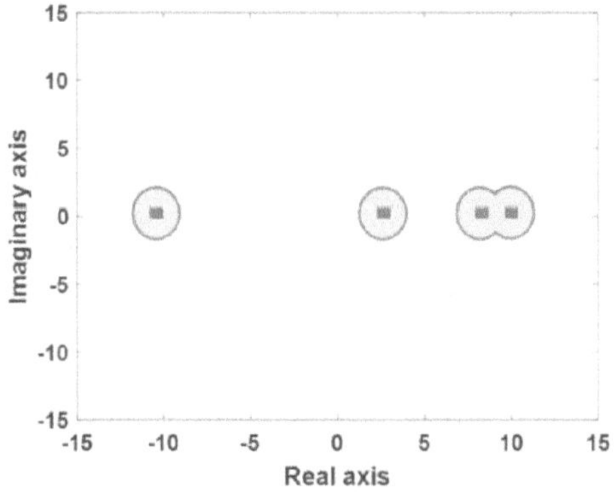

Figure 2.6: The discs in yellow have been derived for eigenvalues. The first two disks overlap and their union contains two eigenvalues. The third and fourth disks are disjoint from the others and contain one eigenvalue each.

the theorem, it can be well understood that every eigenvalue of the matrix $\mathbf{A}$ must be within a distance d of $\mathbf{A}_{ii}$ for some i. Since in general, an eigenvalue can be visualized as a point in the complex plane, it necessitates that the point has to be within distance d of $\mathbf{A}_{ii}$ for some i.

2.10 Time series analysis

Often, data is presented as a time series, which is subsequently modelled. Some of these models are more popular than the others, and an overview follows here. Interested readers can refer to more extensive books [11, 12]. Time series analysis is based on the fundamental idea that it is a realization of the response of a linear, and/or nonlinear system (often represented by a filter) to white noise input. Based on different types of filters, a time series can accommodate a wide variety of structures (or, models) like auto-regressive (AR), autoregressive moving average (ARMA) and non-stationary representations like autoregressive integrated moving average (ARIMA) models.

An autoregressive (AR) model considers a time series y_t expressed as a linear aggregate of past values and an error term z_t which is a white noise process.

$$y_t = \mu + \phi_1 y_{t-1} + \cdots + \phi_p y_{t-p} + z_t \tag{2.39}$$

where y_t is regressed on previous values of itself of order p, with ϕ as model parameters. Using a backshift operator B, this can be written as

$$\phi(B^p)\, y_t \;=\; \mu + z_t \tag{2.40}$$

A moving average (MA) model is a linear filter of order q applied to a white noise process z_t with parameter θ.

$$y_t = \mu + \phi_1 y_{t-1} + \cdots + \phi_p y_{t-p} + z_t \tag{2.41}$$

The combination of an AR model with a MA model produces an autoregressive-moving average (ARMA) model as:

$$y_t = \mu + \phi_1 y_{t-1} + \cdots + \phi_p y_{t-p} + \theta_1 z_{t-1} - \cdots - \theta_q z_{t-q} + z_t \tag{2.42}$$

A nonstationary time series does not have a constant mean and variance and thus, a stationarity transformation is needed. Towards this, y_t can be differenced, leading to an AutoRegressive Integrated Moving Average (ARIMA) model as:

$$y'_t = \mu + \phi_1 y'_{t-1} + \cdots + \phi_p y'_{t-p} + \theta_1 z_{t-1} - \cdots - \theta_q z_{t-q} + z_t \tag{2.43}$$

where y'_t is the differenced series.

In order to estimate the parameters of a time series, the error term can be suitably rephrased to express the model as a least square regression model of the form

$$y_t = \beta_0 + \beta_1 x_{1,t} + \cdots + \beta_r x_(r,t) + \tilde{e_t} \tag{2.44}$$

where y_t is a linear function of the r predictor variables $(x_{1,t}, \ldots, x_{r,t})$ and $\tilde{e_t}$ is an uncorrelated error term which is independent and identically normally distributed (i.i.d.). Of many estimation techniques, the two most popular are least squares regression and maximum likelihood [11]. Alternatively, from the viewpoint of the random process models, the parameters of a time series model can also be estimated by formulating the auto-correlation equations and solving them in a linear equations format using the matrix of autocorrelations and the unknown coefficients, often referred to as Yule Walker method [3].

2.11 Summary

The chapter first presents the ideas of deterministic and random signals and demonstrates how statistical estimates are linked to the understanding of random signals. Subsequently, the Fourier Transform of a signal and related properties, including some aspects of windowing are presented in the frequency domain, which segues into frequency response and impulse response functions linking outputs and inputs with the vibrating system. The power spectral density is then presented next. This completes some of the more fundamental and popular concepts that contribute to the approach taken by the book. The second half of the chapter delves into the core idea of an eigen-perturbation technique and the development of robust indicators of features of interest from it, based on the idea of recursive estimations of covariance. The sensitivity of the estimated eigenvalues is taken up next with relevant theorems. Finally, for the purpose of developing relevant and robust indicators of features of interest for real-time detection, some of the fundamental equations in time series analysis are presented.

References

[1] Kreyszig, E. (2009). Advanced Engineering Mathematics 10th Edition.

[2] Newland, D. E. (2012). An introduction to random vibrations, spectral & wavelet analysis. Courier Corporation.

[3] Hayes, M. H. (2009). Statistical Digital Signal Processing and Modeling. John Wiley & Sons.

[4] Rayleigh, J. W. S. (1894). The Theory of Sound: in Two volumes. 1. Macmillan.

[5] Shin, K. and Hammond, J. (2008). Fundamentals of Signal Processing for Sound and Vibration Engineers. John Wiley & Sons.

[6] Löwdin, P. O. (1962). Studies in perturbation theory. IV. Solution of eigenvalue problem by projection operator formalism. Journal of Mathematical Physics, 3(5): 969–982.

[7] Stewart, Gilbert W. (1998). Perturbation theory for the singular value decomposition.

[8] Krishnan, M., Bhowmik, B., Hazra, B. et al. (2018). Real time damage detection using recursive principal components and time varying auto-regressive modeling. Mechanical Systems and Signal Processing, 101: 549–574.

[9] Tripura, T., Bhowmik, B., Pakrashi, V. et al. (2019). Real-time damage detection of degrading systems. Structural Health Monitoring, 2019: 1475921719861801.

[10] Bhowmik, B., Krishnan, M., Hazra, B. et al. (2018). Real-time unified single-and multi-channel structural damage detection using recursive singular spectrum analysis. Structural Health Monitoring, 2018: 1475921718760483.

[11] Box, G. E., Jenkins, G. M., Reinsel, G. C. and Ljung, G. M. (2015). Time Series Analysis: Forecasting and Control. John Wiley & Sons.

[12] Durbin, J. and Koopman, S. J. (2012). Time Series Analysis by State Space Methods. Oxford University Press.

[13] Solomon, Jr. Otis, M. 1991. PSD computations using Welch's method. [Power Spectral Density (PSD)]. *In*: Sandia National Labs, Albuquerque, NM (United States).

Chapter 3

Single-sensor Real Time Damage Detection Techniques: RSSA and its Variants

3.1 Introduction

Health monitoring of engineering structures subjected to dynamic loads is a subject of key interest in the field of civil and mechanical vibration. The damage induced in the structures due to ground motions [1], strong winds [2] and wave impacts during tidal events and tsunamis [3] causes dynamic changes in real time which if not monitored lead to failure of such systems. Therefore, real time health monitoring is needed to take preventive measures against the total collapse of the structure. Recent years have witnessed enormous advancements in the field of real time structural health monitoring (SHM) in terms of development of algorithms towards damage detection [4–8], modal identification [9, 10] and control [11]. However, existing techniques often neglect the effect of operational conditions in combination with external loads on the damage detection of structural systems. Damage detection in the presence of operational conditions remains a challenging area more so, if approached from a real time perspective and it holds a lot of promise in futuristic development in the context of real time SHM. In real life scenarios, some changes in dynamical characteristics could be caused by

Table 3.1: List of Acronyms.

AR	Auto-Regressive
BW	Bouc-Wen
DOF	Degree of Freedom
DOFs	Degrees of Freedom
DSF	Damage Sensitive Features
EMD	Empirical Mode Decomposition
FFT	Fast Fourier Transform
HHT	Hilbert-Huang Transform
IMF	Intrinsic Mode Function
LNM	Linear Normal Mode
MD	Mahalanobis Distance
MDOF	Multi Degrees of Freedom
MSSA	Multi-channel Singular Spectrum Analysis
FOEP	First Order Eigen Perturbation
PCA	Principal Component Analysis
POV	Proper Orthogonal value
POC	Principal Orthogonal Component
POM	Proper Orthogonal Mode
RDSF	Recursive Damage Sensitive Features
RMD	Recursive Mahalanobis Distance
RMSSA	Recursive Multi-channel Singular Spectrum Analysis
RPCA	Recursive Principal Component Analysis
SDE	Stochastic Differential Equation
SNR	Signal to Noise Ratio
SSA	Singular Spectrum Analysis
SHM	Structural Health Monitoring
SVD	Singular Value Decomposition
TVAR	Time Varying Auto-Regressive

operational or environmental conditions [12–14] which masks the subtle change in signal properties [15], which are caused by damage. If these are not taken into account, the detectability of real time damage detection algorithms becomes questionable. The operational conditions entail change in ambient temperature, loading and induced line noise from instruments in the vicinity. In particular one of the operational conditions, *line noise* has the potential to hamper the damage detection capability of the algorithms [16]. In this chapter, a first order eigenperturbation (FOEP) based hybrid RMSSA-RPCA algorithm is proposed for real time damage detection in the presence of line noise.

In the field of SHM the implementation of data driven statistical methods towards multivariate statistics, signal processing, data analysis among others, have gained significant popularity in damage detection and modal identification case studies. Of several popular multivariate statistical analysis tools, principal component analysis (PCA) is extremely popular in extraction of features from data [17], noise removal, dimensional reduction [18] and others. PCA is focused on finding the orthogonal projection of the data onto a lower dimensional linear space, known as the *principal subspace* which has the maximum possible variance in order to find a hidden linear correlation [6]. However as it is an offline technique operating on batch data it is not applicable to real time damage detection. This motivated the development of FOEP based algorithm Recursive Principal component Analysis (RPCA) [5, 8] which accounts for the continuous streaming of data, proving its applicability for real time damage detection studies in structural systems. However, one of its main limitations lies in dealing with single channel data. This limitation is very elegantly addressed by a single channeled algorithm called singular spectrum analysis (SSA), which is very similar to PCA.

The concept of SSA dates back to late 90s with the first scientific publication of Broomhead [19] relating to the extraction of qualitative information from experimental time series. Owing to the capability of SSA to decompose, separate and reconstruct any time series its application is as diverse as from mathematics, physics and economics all the way to meteorology, oceanology and biomedical science. In recent years SSA is finding application in damage detection, system identification, missing data recovery and forecasting [20]. The drawbacks of SSA can be attributed to two main points, first being the use of batch data which hinders its applicability towards real time health monitoring motivating the development of recursive singular spectrum analysis (RSSA) [4]. Second, it takes uni-variate data into consideration for the analysis which makes it less efficient for health monitoring of multiple degrees of freedom (MDOF) systems. This is overcome by the development of multi-channel singular spectrum analysis (MSSA) [4] which can take batch data from multiple sensors for health monitoring. However as it works using batch data only, it is not amenable to real time health monitoring. This is overcome by an algorithm recursive multi channel singular spectrum analysis (RMSSA), an extension of RSSA which considers real time data streaming in from multiple sensors.

In the recent years the use of TVAR modeling is widely accepted in the area of real time damage detection studies of dynamical systems in the publications [4–6, 8]. TVAR estimates the coefficients in real time which measure the change in the sub-space at each time instant and indicate the same as damage at that instant. One of the premises of multivariate analysis is the measurement of separation between two clusters [15]. In multivariate analysis the use of Mahalanobis distance (MD) as a statistical measure has found numerous applications such as: identification of outliers [21], determination of representativity between parent and sampled data sets [22], in pattern recognition problems viz. k-Nearest Neighbor method (kNN) [23], discriminant analysis [24], disjoint modeling techniques for recognition of distribution [25] and more. The use of Mahalanobis distance however depends on clustering of data and thus not applicable to online implementation. In this regard a recently developed DSF RMD [7], the recursive version of MD is implemented for damage detection case studies in the present work.

Towards addressing the problem of damage identification in the presence of operational noise associated with real time SHM, this chapter addresses the following imporatant issues: *First*, a mathematically consistent framework for RMSSA to account for the real time multivariate time series analysis. *Second*, RMSSA is unified with a RPCA algorithm in a recursive framework for real time noise separation and simultaneous structural damage identification in the presence of operational noise that has shown successful identification of damage without the requirement of preprocessing. *Third*, the RMSSA-RPCA algorithm performs eigen value decomposition (EVD) on the response block Hankel co-variance matrix at each instant of time and obtains modal parameters thereby reducing the time complexity. In conjunction with RDSFs, the RMSSA-RPCA algorithm is successful in identifying structural damage of the order of about 15%. Without the requirement of any special statistical signal processing tool on the physical data, identification to such an extent of damage is a novel contribution in real time SHM. *Finally*, the utility of the RMSSA-RPCA algorithm is extended towards real time damage detection of engineering structures subjected to non-stationary excitation in the presence of operational noise which has shown effective detectability of damage instants.

3.2 Background

3.2.1 Singular Spectrum Analysis (SSA)

As a signal processing tool, the main objective of SSA is to decompose a time series into oscillatory (or harmonic) components and noise, where the harmonic components are either pure or amplitude modulated. SSA comprises of two complimentary stages: decomposition and reconstruction of time series. In the first stage, the signal is decomposed and in the second stage the same is reconstructed excluding the unwanted components, which implements the idea of '*separability*'. The basic methodology of the SSA filtering can be summarized in the following steps:

Step 1: *Time series decomposition*: SSA considers one dimensional data taken from a sensor and decomposes it into a multi-dimensional series which is commonly referred as *trajectory matrix* [4, 26]. The components of the trajectory matrix with a selected windowed length d, also known as embedding dimension provides a trajectory matrix which is commonly known as Hankel matrix (all elements in skew diagonal terms are equal).

Step 2: *EVD on the Trajectory matrix*: By taking the co-variance of the trajectory matrix and decomposing it using eigen value decomposition (EVD), eigenvectors and eigenvalues can be obtained which can successively provide PCs (PCs are decoupled data sets in a projected subspace). The EVD of the co-variance matrix is of the form: $\mathbf{Y} = \mathbf{V\Sigma V^T}$, where $\mathbf{\Sigma}$ is a diagonal matrix of eigen values in descending order and $\mathbf{V}$ is an eigen vector arranged corresponding to eigenvalues.

Step 3: *Grouping*: Grouping is done by appropriate elimination and retaining the unwanted principal components formed from the decomposition of the trajectory matrix in the last step. For the selection of unwanted components for truncation a criterion needs to be set apriori. Assuming the partitioning of the original *embedding* set $S \in [1, 2, \ldots, d]$ into 'm' components results in new grouping dimensions. The Hankel matrix can be reconstructed using such d number rank one elementary matrices [4]. The Hankel matrix thus reconstructed is an approximate estimate of the original consisting of significant and noise removed components.

Step 4: *Time series reconstruction*: Once it is constructed the approximated time series can be constructed through diagonal averaging of the approximated Hankel matrix. The detailed formulation of time series reconstruction is well documented in the literature [4, 20, 26].

The aforementioned formulations provide an idea of the methodology of the SSA algorithm. Two key notes at this point can be noted: (i) the SSA essentially requires a periodic windowing of data and (ii) it is primarily applicable to signals arising from single sensors. These drawbacks impart a constriction on the implementation of SSA for online and multivariate applications.

3.2.2 Multichannel Singular Spectrum Analysis (MSSA)

In the recent years, SSA has emerged as a powerful statistical signal processing tool [27, 28] owing to its wide range of applications in time series analysis towards filtering, recognition and extraction of principal patterns. However, for effective and efficient multivariate data analysis the mathematical structure of SSA alone is not sufficient especially in cases of higher DOF structural systems (e.g., UCLA Factor Building, [4]). On the other hand, MSSA employs the mathematical structure of SSA to formulate a promising multivariate time series analysis tool that potentially accounts for most of the state variables of the coupled systems in the construction of an augmented trajectory matrix [27, 28].

Consider a multivariate signal, $\mathbf{X} = [X_1, X_2, ..., X_L]^T$ arising from L channels with each signal $X_i = [x_1^i, x_2^i, ..., x_N^i]$. MSSA performs *time series composition* on each of the signals resulting in an augmented trajectory matrix padded by decomposition of all channels. However, in this case the dimension of the trajectory matrix becomes $[(d \times L), D]$ since each of the L channels is decomposed into a d dimensional phase space. The structure of the block trajectory matrix takes the form: $[\tilde{\mathbf{Y}}] = [\mathbf{Y}_1, \mathbf{Y}_2, \ldots, \mathbf{Y}_L]^T$. On expansion of the block structure of $[\tilde{\mathbf{Y}}]$ one can get the following form:

$$[\tilde{\mathbf{Y}}] = \left[\begin{pmatrix} x_1^1 & x_2^1 & \cdots & x_d^1 \\ x_2^1 & x_3^1 & \cdots & x_{d+1}^1 \\ x_3^1 & x_4^1 & \cdots & x_{d+2}^1 \\ \vdots & \vdots & \ddots & \vdots \\ x_D^1 & x_{D+1}^1 & \cdots & x_N^1 \end{pmatrix} \begin{pmatrix} x_1^2 & x_2^2 & \cdots & x_d^2 \\ x_2^2 & x_3^2 & \cdots & x_{d+1}^2 \\ x_3^2 & x_4^2 & \cdots & x_{d+2}^2 \\ \vdots & \vdots & \ddots & \vdots \\ x_D^2 & x_{D+1}^2 & \cdots & x_N^2 \end{pmatrix} \cdots \begin{pmatrix} x_1^L & x_2^L & \cdots & x_d^L \\ x_2^L & x_3^L & \cdots & x_{d+1}^L \\ x_3^L & x_4^L & \cdots & x_{d+2}^L \\ \vdots & \vdots & \ddots & \vdots \\ x_D^L & x_{D+1}^L & \cdots & x_N^L \end{pmatrix} \right]^T \tag{3.1}$$

Following steps similar to that of the basic SSA algorithm the MSSA performs EVD on the co-variance matrix of each of the individual Hankel matrices of the augmented trajectory matrix $[\tilde{\mathbf{Y}}]$, yielding L sets of $d \times d$ eigenvalue and eigenvector matrices:

$$[\tilde{\mathbf{Y}}] = [\mathbf{Q}]_{(L\times d)\times(L\times d)} [\mathbf{A}]_{(L\times d)\times(L\times d)} [\mathbf{Q}]^{\mathbf{T}}{}_{(L\times d)\times(L\times d)} \tag{3.2}$$

where $\mathbf{Q}$, $\mathbf{A}$ and D are the block eigenvector matrix, block eigenvalue matrix and block unitary matrix given as:

$$[\mathbf{Q}] = \begin{bmatrix} \mathbf{V}_1 & 0 & \cdots & 0 \\ 0 & \mathbf{V}_2 & \cdots & 0 \\ \vdots & \vdots & \ddots & 0 \\ 0 & 0 & \cdots & \mathbf{V}_L \end{bmatrix}; [\mathbf{A}] = \begin{bmatrix} \mathbf{\Sigma}_1 & 0 & \cdots & 0 \\ 0 & \mathbf{\Sigma}_2 & \cdots & 0 \\ \vdots & \vdots & \ddots & 0 \\ 0 & 0 & \cdots & \mathbf{\Sigma}_L \end{bmatrix} \tag{3.3}$$

It is to be noted that $\mathbf{V}_j$ and $\mathbf{\Sigma}_j$ has a dimension of $d \times d$ and are the individual eigenmatrices of jth channel with $[j \rightarrow 1, 2, 3, \ldots, L]$ channels. The eigenvalue and eigenvector matrices decouples the trajectory matrix through projection of the Hankel matrix onto the eigenvectors resulting in a lower order uncorrelated time series. The process of de-correlation obtains L sets of d–PCs, each set indicating the optimal subspace of the trajectory matrices. The block matrix $\tilde{\mathbf{Z}}$ containing individual PCs can be expressed as:

$$\tilde{\mathbf{Z}} = \begin{bmatrix} \mathbf{Z}_1 \\ \mathbf{Z}_2 \\ \vdots \\ \mathbf{Z}_L \end{bmatrix} = \begin{bmatrix} \mathbf{V}_1^T & 0 & \cdots & 0 \\ 0 & \mathbf{V}_2^T & \cdots & 0 \\ \vdots & \vdots & \ddots & \vdots \\ 0 & 0 & 0 & \mathbf{V}_L^T \end{bmatrix} \begin{bmatrix} \mathbf{Y}_1 \\ \mathbf{Y}_2 \\ \vdots \\ \mathbf{Y}_L \end{bmatrix} \tag{3.4}$$

For *grouping*, the PCs can be selected to avoid any noise component in signal, by accounting for only the significant energy contents. In this context, it can be understood that the choices of PCs for different channels may differ from others. However, for ease of understanding assuming that m PCs actively participates in construction of elementary matrices, *grouping* results in reduction of embedding dimension from d to m for all the channels. The grouped Hankel matrix can be reconstructed using selected PCs and elementary matrices as follows:

$$\tilde{\mathbf{Y}} = \begin{bmatrix} \sum \mathbf{R}_i^1 \\ \sum \mathbf{R}_i^2 \\ \vdots \\ \sum \mathbf{R}_i^L \end{bmatrix} = \begin{bmatrix} \sum V_i^1 \times Z_i^1 \\ \sum V_i^2 \times Z_i^2 \\ \vdots \\ \sum V_i^L \times Z_i^L \end{bmatrix} = \sum_{i=1}^{m} \begin{bmatrix} V_i^1 & 0 & \cdots & 0 \\ 0 & V_i^2 & \cdots & 0 \\ \vdots & \vdots & \ddots & \vdots \\ 0 & 0 & \cdots & V_i^L \end{bmatrix} \begin{bmatrix} Z_i^1 \\ Z_i^2 \\ \vdots \\ Z_i^L \end{bmatrix} \tag{3.5}$$

Once the Hankel matrix is reconstructed the original multivariate signal $\mathbf{X} = [X_1, X_2, ..., X_L]^T$ can be obtained from the trajectory matrix by following same diagonal averaging as explained in step 4 of Filtering. Thus MSSA algorithm solves the problem of multivariate filtering as a sequal to the univariate SSA filtering. Although MSSA can perform multivariate filtering it still operates in batch mode. Thus a complete reformulation is necessary for online implementation.

The drawbacks associated with the aforementioned SSA and MSSA algorithms can be mainly attributed to periodic windowing of data and requirement of healthy system responses rendering them inoperable in online mode. In this context, online implementation of the aforementioned algorithms is possible through the estimation of eigenvector and eigenvalues at each time instant. This is possible by performing EVD on the Hankel co-variance matrix at each time instant. However without any special formulation, straightforward application of EVD at each instant of time may result in time complexity and memory exhaustion since the co-variance data matrix grows with streaming data. These shortcomings can be overcome through the use of first order eigenperturbation approach (FOEP) which provides recursive updates of eigenspace at each instant of time by utilizing the eigenspace at previous instant that can be directly related to the eigenvalues and eigenvectors [4,29]. On a similair note, inspite of the advantages of MSSA over SSA due to accessibility of multichannel time series, it becomes computationally more involved due to treatment of EVD on augmented Hankel co-variance matrix at each time step. The development of recursive counterpart of MSSA by modifying the basic formulation of MSSA algorithm using the principles of FOEP technique to account for real time analysis of multiple input data is shown in this chapter. This not only solves the problem of time complexity and mathematical consistency but also processes data as soon as it streams in.

3.2.3 RPCA in structural dynamics

The theoretical development of RPCA premises on the real time implementation of traditional PCA through FOEP based framework. RPCA is an orthogonal transform which obtains the POMs and POVs from the physical multivariate data by preforming the EVD on sample co-variance matrix assuming a new data is recorded at every instant of time. The orthogonal transformation has the form $\mathbf{Z} = \mathbf{VX}$, such that the POMs closely emulates the theoretical linear normal modes (LNMs) (i.e., mode-shapes of vibrating system). The correlation between POMs from RPCA and theoretical LNMs in the light of structural dynamics can be understood by considering the generous dynamical system of the form:

$$\mathbf{MX}(t) + \mathbf{CX}(t) + \mathbf{KX}(t) = \mathbf{F}(t) \tag{3.6}$$

where mass, stiffness and damping matrices are represented by **M**, **C** and **K** with X as the displacement vector and $\mathbf{F}(t)$ as the input excitation. The solution of the equation can be written as $\mathbf{X} = \mathbf{VQ}$, where **X** is the measurement matrix of size $m \times N$ and **Q** is the corresponding modal matrix of size $m \times N$ with m as number of degrees of freedom and N as the sampling size. **V** is a orthogonal transformation matrix of size $m \times m$ yielding mode matrix such that the mode shape matrix are orthogonal to each other with respect to the matrix **M**. Through the modal transformation, the co-variance matrix $\mathbf{R}_X = \frac{1}{N}\mathbf{XX}^T$ can be expressed as $\mathbf{R}_X = \frac{1}{N}\mathbf{VQQ}^T\mathbf{V}^T$, where $\mathbf{R}_\mathbf{Q} = \frac{1}{N}\mathbf{QQ}^T$ represents the co-variance matrix of the modal responses. Since the matrices **M**, **C** and **K** are diagonalized in the modal coordinate the modal responses ($\mathbf{Q} = [q_1, q_2, \ldots q_i \ldots q_n]^T$) of the Eq. 5.1 can be written as:

$$\ddot{q}_i(t) + 2\zeta_i\omega_{n,i}\dot{q}_i(t) + \omega^2{}_{n,i}q_i(t) = \frac{1}{m_i}f_i(t) \tag{3.7}$$

where, $f_i(t) = v_i{}^T\mathbf{F}(t)$ is the modal force corresponding to the v_i mode. ζ_i, $\omega^2{}_{n,i}$ and m_i are the modal damping, modal frequency and modal mass for i^{th} mode. The solution of the Eq. 3.7 is obtained by using Duhamel Integral: $q_i(t) = \int_0^\infty f_i(\tau)h_i(t-\tau)\,d\tau$, where $h_i(t-\tau) = \frac{1}{m_i\,\omega_{d,i}}e^{-\zeta_i\,\omega_{n,i}(t-\tau)}\sin\,\omega_{d,i}(t-\tau)$ and $\omega^2{}_{d,i}$ is the damped modal frequency. The solution matrix **Q** in the vectorial form can be expressed as (using convolution property):

$$\mathbf{Q}_{i\times 1}(t) = \int_0^\infty \underbrace{\begin{bmatrix} h_1(t-\tau) & 0 & \cdots & 0 \\ 0 & h_2(t-\tau) & \cdots & 0 \\ \vdots & \vdots & \ddots & \vdots \\ 0 & 0 & \cdots & h_i(t-\tau) \end{bmatrix}_{i\times i}}_{\mathbf{H}^Q(t-\tau)} \underbrace{\begin{Bmatrix} f_1(\tau) \\ f_2(\tau) \\ \vdots \\ f_i(\tau) \end{Bmatrix}_{i\times 1}}_{F^Q(\tau)} d\tau \tag{3.8}$$

The correlation matrix for the modal responses, $\mathbf{R_Q}$ can be expressed as linear combination of above entities $\mathbf{H}^Q(t-\tau)$ and $F^Q(\tau)$. The POCs ($\mathcal{P}$) or orthogonal transformation of the data obtained by RPCA analysis can be written as product of POMs ($\mathbf{W}$) and the data vector ($\mathbf{X}$) as in Eq. Equation (5.3). In order to perform further analysis the POCs ($\mathcal{P}$) can be expressed as a sum of true linear normal coordinates (Q) and an error term ($\mathcal{E}$):

$$\begin{aligned}\mathcal{P}(t) &= \mathbf{W}^T\mathbf{X}(t)\\ &= Q(t)+\mathcal{E}\end{aligned} \tag{3.9}$$

Where, $\mathcal{P} = [\psi_1, \psi_2 \ldots \psi_m]^T$, $Q = [q_1, q_2 \ldots q_m]^T$ and $\mathcal{E} = [\varepsilon_1, \varepsilon_2 \ldots \varepsilon_m]^T$, m denoting the size of dynamical system. To understand the behavior of POCs in collation of modal responses it is imperative to construct the co-variance matrix $\mathbf{R}_{\mathcal{P}} = \frac{1}{N}\mathcal{P}\mathcal{P}^{\mathbf{T}}$ as:

$$\begin{aligned}\mathbf{R}_{\mathcal{P}} = \frac{1}{N}&\begin{bmatrix}\sum_{k=1}^{N} q_1(t_k)q_1(t_k) & \sum_{k=1}^{N} q_1(t_k)q_2(t_k) & \cdots & \sum_{k=1}^{N} q_1(t_k)q_i(t_k)\\ \sum_{k=1}^{N} q_2(t_k)q_1(t_k) & \sum_{k=1}^{N} q_2(t_k)q_2(t_k) & \cdots & \sum_{k=1}^{N} q_2(t_k)q_i(t_k)\\ \vdots & \vdots & \ddots & \vdots\\ \sum_{k=1}^{N} q_i(t_k)q_1(t_k) & \sum_{k=1}^{N} q_i(t_k)q_2(t_k) & \cdots & \sum_{k=1}^{N} q_i(t_k)q_i(t_k)\end{bmatrix}_{m\times m} + \frac{1}{N}\begin{bmatrix}\sum_{k=1}^{N} q_1(t_k)\varepsilon_1(t_k) & \sum_{k=1}^{N} q_1(t_k)\varepsilon_2(t_k) & \cdots & \sum_{k=1}^{N} q_1(t_k)\varepsilon_i(t_k)\\ \sum_{k=1}^{N} q_2(t_k)\varepsilon_1(t_k) & \sum_{k=1}^{N} q_2(t_k)\varepsilon_2(t_k) & \cdots & \sum_{k=1}^{N} q_2(t_k)\varepsilon_i(t_k)\\ \vdots & \vdots & \ddots & \vdots\\ \sum_{k=1}^{N} q_i(t_k)\varepsilon_1(t_k) & \sum_{k=1}^{N} q_i(t_k)\varepsilon_2(t_k) & \cdots & \sum_{k=1}^{N} q_i(t_k)\varepsilon_i(t_k)\end{bmatrix}_{m\times m}\\ + \frac{1}{N}&\begin{bmatrix}\sum_{k=1}^{N} \varepsilon_1(t_k)q_1(t_k) & \sum_{k=1}^{N} \varepsilon_1(t_k)q_2(t_k) & \cdots & \sum_{k=1}^{N} \varepsilon_1(t_k)q_i(t_k)\\ \sum_{k=1}^{N} \varepsilon_2(t_k)q_1(t_k) & \sum_{k=1}^{N} \varepsilon_2(t_k)q_2(t_k) & \cdots & \sum_{k=1}^{N} \varepsilon_2(t_k)q_i(t_k)\\ \vdots & \vdots & \ddots & \vdots\\ \sum_{k=1}^{N} \varepsilon_i(t_k)q_1(t_k) & \sum_{k=1}^{N} \varepsilon_i(t_k)q_2(t_k) & \cdots & \sum_{k=1}^{N} \varepsilon_i(t_k)q_i(t_k)\end{bmatrix}_{m\times m} + \frac{1}{N}\begin{bmatrix}\sum_{k=1}^{N} \varepsilon_1(t_k)\varepsilon_1(t_k) & \sum_{k=1}^{N} \varepsilon_1(t_k)\varepsilon_2(t_k) & \cdots & \sum_{k=1}^{N} \varepsilon_1(t_k)\varepsilon_i(t_k)\\ \sum_{k=1}^{N} \varepsilon_2(t_k)\varepsilon_1(t_k) & \sum_{k=1}^{N} \varepsilon_2(t_k)\varepsilon_2(t_k) & \cdots & \sum_{k=1}^{N} \varepsilon_2(t_k)\varepsilon_i(t_k)\\ \vdots & \vdots & \ddots & \vdots\\ \sum_{k=1}^{N} \varepsilon_i(t_k)\varepsilon_1(t_k) & \sum_{k=1}^{N} \varepsilon_i(t_k)\varepsilon_2(t_k) & \cdots & \sum_{k=1}^{N} \varepsilon_i(t_k)\varepsilon_i(t_k)\end{bmatrix}_{m\times m}\end{aligned} \tag{3.10}$$

Equation (3.10) can be written in the concise form as shown below.

$$\begin{aligned}\mathbf{R}_{\mathcal{P}} &= \int_0^{\infty}\int_0^{\infty} \frac{1}{N}\sum_{k=1}^{N} \mathbf{H}^Q_{m\times m}F^Q_{m\times 1}F^{Q^T}_{1\times m}\mathbf{H}^Q_{m\times m}d\tau d\theta + \int_0^{\infty} \frac{1}{N}\sum_{k=1}^{N} \mathbf{H}^Q_{m\times m}F^Q_{m\times 1}\mathcal{E}^T_{1\times m}d\tau +\\ &\int_0^{\infty} \frac{1}{N}\sum_{k=1}^{N} \mathcal{E}_{m\times 1}F^{Q^T}_{1\times m}\mathbf{H}^{Q^T}_{m\times m}d\tau + O(h^2)\end{aligned} \tag{3.11}$$

In practical studies structural systems have a very low to moderate damping percentages (generally 1-3%) by virtue of which the Eq. 7.6 provides a good approximation to the theoretical linear normal modes (LNMs) which deviates as damping increases. Further it can be understood from Eq. 7.6 that as sample size increases ($N \rightarrow \infty$) the error vector $\mathcal{E}$ tends to zero magnitude and POC converges to the linear normal modal coordinates as the POC co-variance matrix $R_{\mathcal{P}}$ becomes diagonally dominant. The main approach is to develop a formulation for RPCA with FOEP based framework that can provide a recursive estimate of updates of POMs, i.e., to the estimate of POC vector: $\mathcal{P}_k$ at each instant of time in a real time framework. At this point it is to be noted that the modal coordinates are uncorrelated and orthogonal in the transformed subspace which reduces the requirement of model dimension required in time series modeling of system responses thereby reducing the degree of memory complexity. Since, the algorithm estimates the updates for eigenspace corresponding to instantaneous co-variance matrices $\mathbf{R}_k$, it is essential to evaluate the co-variance matrix $\mathbf{R}_k$ in terms of the previous time instant, i.e., $\mathbf{R}_{k-1}$ and response vector X_k at the kth time instant as: $\mathbf{R}_k = \frac{k-1}{k}\mathbf{R}_{k-1} + \frac{1}{k}X_kX_k^T$. The RPCA algorithm performs an eigen value decomposition (EVD) on the co-variance estimate $\mathbf{R}_k$ which enables the transformation: $[\mathbf{R}_k] = [\mathbf{W}_k]\,[\mathbf{\Theta}_k]\,[\mathbf{W}_k]^T$. The generalized form of the EVD on $\mathbf{R_k}$ can be expressed as:

$$[\tilde{\mathbf{W}}_k][k\tilde{\mathbf{\Theta}}_k][\tilde{\mathbf{W}}_k]^T = (k-1)\mathbf{\Sigma}_k^{-1}\mathbf{\Sigma}_{k-1}[\tilde{\mathbf{W}}_{k-1}][\tilde{\mathbf{\Theta}}_{k-1}][\tilde{\mathbf{W}}_{k-1}]^T\mathbf{\Sigma}_{k-1}\mathbf{\Sigma}_k^{-1} + \mathbf{\Sigma}_k^{-1}\Delta\mu_k\Delta\mu_k^T\mathbf{\Sigma}_k^{-1} + \frac{1}{k}\,[X_k - \mu_k]\,[X_k - \mu_k]^T \tag{3.12}$$

where the recursive mean is estimated as: $\mu_k = \frac{k-1}{k}\mu_{k-1} + \frac{1}{k}X_k$. However, the structural systems are generally evolved as zero mean process and further scaling of data into unit variance produces: $\tilde{\mathbf{W}}_k = \mathbf{W}_k$. Again, using the relationship of POM and LNM Eq. (5.3), one can write: $X_k = \mathbf{W}_{k-1}\mathcal{P}_k$ which reduces the Eq. 3.12 into following form:

$$[\mathbf{W}_k]\,[k\mathbf{\Theta}_k]\,[\mathbf{W}_k]^T = \mathbf{W}_{k-1}\left[(k-1)\,\mathbf{\Theta}_{k-1} + \mathcal{P}_k\mathcal{P}_k^T\right]\mathbf{W}_{k-1}^T \tag{3.13}$$

When the norm of off-diagonal elements are small compared to the norm of leading diagonal the POMs converges to the LNMs and thus the robustness of the RPCA algorithm significantly improves when the the term $[(k-1)\,\mathbf{\Theta}_{k-1} + \mathcal{P}_k\mathcal{P}_k^T]$ is diagonally dominant [30]. The co-variance of POCs $::\mathcal{P}_k\mathcal{P}_k^T$ is evaluated using Eq. 5.3, scaled to an arbitrary factor:

$$\begin{aligned} \mathcal{P}_k\mathcal{P}_k^T &= \mathbf{W}_{k-1}^T X_k {X_k}^T \mathbf{W}_{k-1} \\ &= Q_{k-1}{Q_{k-1}}^T + \underbrace{\mathcal{E}{Q_{k-1}}^T + \mathcal{E}^T Q_{k-1}}_{1^{st} order\ error} + \underbrace{\mathcal{E}\mathcal{E}^T}_{2^{nd} order\ error} \end{aligned} \tag{3.14}$$

From aforementioned background it is understood that the second order error terms in the Eq. 7.14 becomes negligible under low damping and large number of sampling points [31].

3.3 RMSSA-RPCA adaptive algorithm

To operate in real time for separation of line noise from signal and simultaneous damage detection in a single framework the proposed algorithm primarily comprises of two segments: (i) Filtering and (ii) Damage detection. The Filtering utilizes *Recursive Multichannel Singular Spectrum Analysis* (RMSSA) to filter out the noise components manifested by signal noises, environmental and operational conditions. The mathematical structure of proposed RMSSA algorithm is analogous to traditional *Singular Spectrum Analysis* (SSA) but utilized the principles of first order eigenperturbation technique (FOEP) for real time series analysis, primarily for online separation of unwanted components. The RMSSA algorithm is also an advanced structure of recently established RSSA [4] algorithm with built-in multivariate analysis feature that was lacking in RSSA algorithm therefore can account for multivariate analysis without much computational cost. The RMSSA algorithm produces a set of filtered data series having a lower model order than the original signal. The reconstructed signal has significant information regarding characteristics of source and a better representative of structural damage. The filtered response of RMSSA is taken as input to the **damage detection** that utilizes *Recursive Principal Component Analysis* (RPCA) on the colored signal (filtered) for identification of damage events. The RPCA is the recursive counterpart of *Principal Component Analysis* (PCA) and has shown great damage identification in real time in recently published literature [5,6].

Towards understanding the mathematical formulation RMSSA, consider the Hankel co-variance matrix of Eq. 3.1 of the form: $\mathbf{C}_Y = \frac{1}{N}\mathbf{Y}\mathbf{Y}^T$. For any multivariate data set the recursive estimation of the Hankel co-variance matrix ($\mathbf{C}_k$) at time instant k can be expressed in terms of the data vector at k^{th} instant, i.e., (X_k) as:

$$\mathbf{C}_k = \frac{k-1}{k}\mathbf{C}_{k-1} + \frac{1}{k}\mathbf{Y}_k\mathbf{Y}_k^T \tag{3.15}$$

where $\mathbf{C}_{k-1}$ is the co-variance estimate at $(k-1)^{th}$ instant. The similar concept of recursive co-variance estimation through FOEP techniques can be applied for the real time processing of augmented trajectory matrix in Eq. Equation (3.1), as follows:

$$\begin{bmatrix} \mathbf{C}_k^1 \\ \mathbf{C}_k^2 \\ \vdots \\ \mathbf{C}_k^L \end{bmatrix} = \frac{k-1}{k}\begin{bmatrix} \mathbf{C}_{k-1}^1 \\ \mathbf{C}_{k-1}^2 \\ \vdots \\ \mathbf{C}_{k-1}^L \end{bmatrix} + \frac{1}{k}\begin{bmatrix} \{Y_k^1\}\{Y_k^1\}^T \\ \{Y_k^2\}\{Y_k^2\}^T \\ \vdots \\ \{Y_k^L\}\{Y_k^L\}^T \end{bmatrix} \tag{3.16}$$

where Y_k^i is the trajectory vector expanding from k^{th} sample expressed as, $Y_k^i = \{x_{k-d+1}^i, x_{k-d+2}^i, x_{k-d+3}^i \ldots x_k^i\}$. The primary aim of RMSSA is to estimate the update of augmented co-variance matrix at each time instant without actually performing EVD on the block co-variance matrices thereby reducing the time and memory complexity. This is possible using FOEP techniques that allows the estimation of eigenspace without actually performing the EVD recursively. The individual Hankel co-variance estimates of the augmented trajectory matrix at k^{th} instant can be expressed by its EVD as, $\mathbf{C}_k^j = \mathbf{V}_k^j \mathbf{\Sigma}_k^j \mathbf{V}_k^{j^T}$ where $\mathbf{V}_k^j$ and $\mathbf{\Sigma}_k^j$ represents the eigenvector and eigenvalue matrices for j^{th} channel ($j \in 1, 2, 3, \ldots L$) at k^{th} instant, respectively. For non-stationary processes the drift in mean level can be accommodated in the estimate of co-variance update by using mean shift in the data samples as follows:

$$\mathbf{V}_k^j \mathbf{\Sigma}_k^j \mathbf{V}_k^{j^T} = \frac{(k-1)}{k} \mathbf{V}_{k-1}^j \mathbf{\Sigma}_{k-1}^j \mathbf{V}_{k-1}^{j\ \ T} + \frac{1}{k} \left[Y_k^j - \mu_k^j \right] \left[Y_k^j - \mu_k^j \right]^T \tag{3.17}$$

where, the recursive mean at k^{th} instant is estimated as: $\mu_k^j = \frac{k-1}{k} \mu_{k-1}^j + \frac{1}{k} Y_k^j$. However, for structural systems the response data vectors are generally evolve as a zero mean processes enabling the Eq. 3.17 to be written without any mean shift. Further simplifying the Eq. 3.17 with the relation $Y_k^j = \mathbf{V}_{k-1}^j Z_k^j$ one gets:

$$\mathbf{V}_k^j k \mathbf{\Sigma}_k^j \mathbf{V}_k^{j^T} = \mathbf{V}_{k-1}^j \{ (k-1) \mathbf{\Sigma}_{k-1}^j + Z_k^j Z_k^{j^T} \} \mathbf{V}_{k-1}^{j\ \ T} \tag{3.18}$$

Under finitely large sample-size k the term $\{ (k-1) \mathbf{\Sigma}_{k-1}^j + Z_k^j Z_k^{j^T} \}$ exhibits a diagonally dominant characteristics since $\mathbf{\Sigma}_{k-1}^j$ is diagonally dominant. Towards obtaining a mapping between both the sides of Eq. 3.18, Gershgorin's theorem ensures the EVD of the term to be of the form $\mathbf{\Psi}_k^j \mathbf{\Omega}_k^j \left(\mathbf{\Psi}_k^j \right)^T$, where $\mathbf{\Psi}_k^j$ is the orthonormal eigenvector and $\mathbf{\Omega}_k^j$ is approximated eigenvalue matrices at k^{th} instant. Upon substitution and simplification the Eq. 3.18 can be rewritten as follows:

$$\mathbf{V}_k^j k \mathbf{\Sigma}_k^j \mathbf{V}_k^{j^T} = \mathbf{V}_{k-1}^j \mathbf{\Psi}_k^j \mathbf{\Omega}_k^j \left(\mathbf{\Psi}_k^j \right)^T \mathbf{V}_{k-1}^{j\ \ T} \tag{3.19}$$

On close observation of the Eq. 6.14, final updates for the eigenvector and eigenvalues can be identified as:

$$\begin{bmatrix} \mathbf{V}_k^1 \\ \mathbf{V}_k^2 \\ \vdots \\ \mathbf{V}_k^L \end{bmatrix} = \begin{bmatrix} \mathbf{V}_{k-1}^1 & 0 & \cdots & 0 \\ 0 & \mathbf{V}_{k-1}^2 & \cdots & 0 \\ \vdots & \vdots & \ddots & \vdots \\ 0 & 0 & \cdots & \mathbf{V}_{k-1}^L \end{bmatrix} \begin{bmatrix} \mathbf{\Psi}_k^1 \\ \mathbf{\Psi}_k^2 \\ \vdots \\ \mathbf{\Psi}_k^L \end{bmatrix} \quad \& \quad \begin{bmatrix} \mathbf{\Sigma}_k^1 \\ \mathbf{\Sigma}_k^2 \\ \vdots \\ \mathbf{\Sigma}_k^L \end{bmatrix} = \frac{1}{k} \begin{bmatrix} \mathbf{\Omega}_k^1 \\ \mathbf{\Omega}_k^2 \\ \vdots \\ \mathbf{\Omega}_k^L \end{bmatrix} \tag{3.20}$$

A drawback for using FOEP technique to estimate the eigenspace at each instant is associated with permutation ambiguity, i.e., the recursive eigenvectors and eigenvalues are not arranged in same sequence as the theoretical values. However the shortcoming can be overcome by rearranging the eigenvalues in decreasing order, and, correspondingly the eigenvectors. Following Eq. 3.4, the principal components of each trajectory matrix in $\mathbf{Y}^j$ at a particular time instant can be extracted as follows:

$$\mathbf{Z}^j(k) = \mathbf{\Psi}^{j^T}(k)\mathbf{Y}^j(k); j \in 1,2,3,\ldots L \tag{3.21}$$

At this point rephrasing the fact that each of the L channels are decomposed into d trajectories, it is assumed that out of d PCs, m contribute to the reconstruction of original signal. For j^{th} channels the corresponding elementary vectors $\mathbf{R}_i^j(k); i \in 1,2,\ldots m$ can be constructed by projecting the retained PCs back into its original subspace as previously mentioned:

$$\mathbf{Y}^j(k) = \sum_{i=1}^{m} \mathbf{R}_i^j(k) = \sum_{i=1}^{m} \mathbf{V}_i^j(k)_{m\times 1} \times \mathbf{Z}_i^j(k) \tag{3.22}$$

Proper mathematical operation for i^{th} trajectory and j^{th} channel the construction of augmented trajectory vector $[\tilde{\mathbf{Y}}(k)] = [\mathbf{Y}^1(k), \mathbf{Y}^2(k), \ldots, \mathbf{Y}^L]^T(k)$ is straightforward. It is to be noted that the degree of correlation of the reconstructed trajectory matrix with the original depends on the removal of PCs. Finally at each instant the original sequence $\mathbf{U}(k) = [U^1(k), U^2(k), ..., U^L(k)]^T$ can be derived from the augmented Hankel matrix $[\tilde{\mathbf{Y}}]$ by simple averaging as shown below:

$$U^j(k) = \frac{1}{m}\sum_{i=1}^{m} \mathbf{Y}_i^j(k) \tag{3.23}$$

At the end of this step the obtained signals are filtered and noise removed which are responses of reduced order and contains most of the information about alteration in the subspace thereby indicating damage in structure. Once the original signal at each time instant is obtained, the RPCA module utilizes the response vector at each instant to identify and detect the damage.

The RPCA module obtains the co-variance matrix of the filtered responses and evaluates the EVD at each successive instant using FOEP technique. The EVD on covaiance yields eigenspace consisting of eigenvectors and eigenvalues which provides reduced order transformed responses which can be expressed in similar steps as Eq. 3.12 and Eq. 5.8. On observation of Eq. 7.14 it can be noted that the first term resembles the co-variance of normal coordinates at $(k-1)^{th}$ instant, and rest of the terms are representative of first order errors. Since the normal coordinates are uncorrelated in transformed space the entity $Q_{k-1}{Q_{k-1}}^T$ can be approximated as diagonally dominant whose EVD is close to the EVD of its diagonal entries. To ensure robust diagonal dominance of the matrix $\left[(k-1)\mathbf{\Theta}_{k-1} + \mathcal{P}_k\mathcal{P}_k^T\right]$, the first order errors $\mathcal{E}{Q_{k-1}}^T$ and $\mathcal{E}^T Q_{k-1}$ needs to eliminated from the POC co-variance estimate $\mathcal{P}_k\mathcal{P}_k^T$, enabling direct application of Gershgorin's theorem [30]. A feedback of the form $\Lambda\Lambda^T = \mathcal{E}{Q_{k-1}}^T + \mathcal{E}^T Q_{k-1}$

accounting for the error terms can be utilized to obtain the modal vectors at each instant k^{th} from POCs as:

$$[\mathbf{W}_k]\,[k\boldsymbol{\Theta}_k]\,[\mathbf{W}_k]^T = \mathbf{W}_{k-1}\left[(k-1)\,\boldsymbol{\Theta}_{k-1} + \mathcal{P}_{k-1}\mathcal{P}_{k-1}{}^T - \mathcal{E}\mathcal{Q}_{k-1}{}^T - \mathcal{E}^T\mathcal{Q}_{k-1}\right]\mathbf{W}_{k-1}^T$$
$$= \mathbf{W}_{k-1}\left[(k-1)\,\boldsymbol{\Theta}_{k-1} + \underbrace{\mathcal{Q}_{k-1}\mathcal{Q}_{k-1}{}^T}_{modal\ Cov}\right]\mathbf{W}_{k-1}^T \tag{3.24}$$

The update equations of the EVD of the matrix $\mathbf{W}_{k-1}\left[(k-1)\,\boldsymbol{\Theta}_{k-1} + \mathcal{Q}_{k-1}\mathcal{Q}_{k-1}{}^T\right]\mathbf{W}_{k-1}^T$ can be substituted as $\boldsymbol{\Lambda}_k\mathbf{D}_k\boldsymbol{\Lambda}_k^T$ into Eq. 5.8 as, $[\mathbf{W}_k][k\boldsymbol{\Upsilon}_k][\mathbf{W}_k]^T = [\mathbf{W}_{k-1}\boldsymbol{\Lambda}_k][\mathbf{D}_k][\mathbf{W}_{k-1}\boldsymbol{\Lambda}_k]^T$, which yields the following iterative update equations:

$$\left.\mathbf{W}_k = \mathbf{W}_{k-1}\boldsymbol{\Lambda}_k \qquad \& \qquad \boldsymbol{\Upsilon}_k = \frac{\mathbf{D}_k}{k}\right\} \tag{3.25}$$

It is evident from Eq. 5.11 that the recursive estimate of $\mathbf{R}_k$ using a FOEP based framework requires the evaluation of the values, Λ_k and D_k. With the correction: $\mathcal{Q}_{k-1}\mathcal{Q}_{k-1}{}^T = \mathcal{P}_k\mathcal{P}_k{}^T - \Lambda\Lambda^T$ the term $[(k-1)\,\boldsymbol{\Theta}_{k-1} + \mathcal{P}_k\mathcal{P}_k^T]$ becomes diagonally dominant, and the eigenvalues can now be assumed to be equivalent of the eigenvalues of diagonal entries of the matrix. The error can be estimated by evaluating the Euclidean norm between true LNM and POM obtained from the initial batch population. Recently published literature [8] has shown that in the presence of such errors the POMs obtained using RPCA does not converge to true LNM even the size of sample population increases.

To understand the approximation of LNMs for the vibrating modes through the incorporation of errors, a simple case study using a 3 DOF linear system having identical floors of mass 2kg and linear story stiffness of 1kN/m is undertaken [8]. The structural damping at each floor is assumed as 2% of critical damping ratio. The present work assumes an error of approximately 2%, 6% and 14% expressed against respective LNMs [8]. Case studies for higher DOFs reveals that the mass participation of higher modes is comparatively less than other modes thus consideration of errors for higher modes doesn't have much effect. It is to be noted that the estimate of errors is carried out using a batch of few initial data samples since the information on the error estimate is essential for the real time framework. The approximated LNMs provides better information on the convergence of the vibration modes as data streams in real time thereby facilitating more accurate damage identification specially in presence of operational noises where the sensitivity for operational factors is a prime concern.

Using the aforementioned concepts, the ith diagonal entry of eigenvalues, D_k can be represented as $(k-1)\,\lambda_i + \Psi_i^2$, where λ_i is the $(i,i)^{th}$ element of $\boldsymbol{\Theta}_{k-1}$

and Ψ_i is the ith entry of the POC estimate. Similarly the corresponding value of $\mathbf{\Lambda}_k$ can be found out using the relation: $\mathbf{\Lambda}_k = (I + \Delta_\Lambda)$, where $(i, j)^{th}$ element of Δ_Λ is $(\psi_i \psi_j)/(\upsilon_j + \psi_j^2 - \upsilon_i - \psi_i^2)$ and $(i, j)^{th}$ elements are 0. The recursive eigenvectors obtained at each time instant are not ordered and hence need to be arranged based on the decreasing order of the corresponding eigenvalues in Θ_k resorting to the resolution of permutation ambiguity [32]. The contribution factor of eigenvector W_r corresponding to r^{th} eigenvalue at each time instant can be written as $\alpha_r^2/(\sum_{r=1}^{n} \alpha_i r^2)$, where α_r^2 is the eigenvalue corresponding to W_r [5,6]. The POC vectors at each instant are now utilized for modeling the recursive damage sensitive features.

3.4 Framework of the RMSSA-RPCA algorithm

This section summarizes the basic methodology of the RMSSA-RPCA algorithm towards real time damage detection of structural systems in the presence of operational noise. The working principal of the algorithm premises on three primary segments operating *simultaneously* in a recursive framework: Firstly, removal of operational noise from the online streaming raw data by RMSSA, Secondly, processing of the filtered response from RMSSA through RPCA to generate transformed data sets at each instant and Finally, employment of the RDSFs (viz., TVAR and RMD) on transformed responses to check for any damage an its subsequent detection. The basic steps of the algorithm in short is captured in Algorithm 1.

The outlines of the proposed algorithm is illustrated in Fig. 3.1. Few key entailment of the proposed algorithm can be listed as: (i) The proposed algorithm processes data without pre-windowing in batches assuming that new data is recorded at each instant of time that makes the algorithm online, (ii) The algorithm operates without the need of historical data it can be considered as less baseline reliant algorithm. However, since the algorithm compares the instances before and after k^{th} data to identify damage the assumption of baseline free does not strictly hold, (iii) The algorithm is model and parameter free technique and can be applied to any of the signal processing field wherever the mathematical structure of the problem facilitates eigen value decomposition, (iv) The algorithm demonstrates its significance in separation of line noises in real time with real time damage detection in a simultaneous framework which is a key novelty in the field of real time structural health monitoring.

Algorithm 1 RMSSA-RPCA algorithm

Require: $\mathbf{X} = [X_1, X_2, ..., X_L]^T \forall X \in \Re^n$ ▹ Response vector of system

for $k = 1 : N_{iteration}$ **do** ▹ Iteration for sample length
  → *Beginning of RMSSA module :*
  $\tau \leftarrow$ lag, $d \leftarrow$embidding dimension
  for $i = 1 : DOF$ **do** ▹ Iteration for Multi-channel
    Estimate Initial Hankel Matrix: $[\tilde{\mathbf{Y}}]$ ▹ Eq. 3.1
    Recursive update of Hankel Covariance matrix ▹ Eq. 3.16
    $\mathbf{V}^j_k k \boldsymbol{\Sigma}^j_k \mathbf{V}^{j\,T}_k = \mathbf{V}^j_{k-1} \boldsymbol{\Psi}^j_k \boldsymbol{\Omega}^j_k \left(\boldsymbol{\Psi}^j_k\right)^T \mathbf{V}^{j\,T}_{k-1}$ ▹ Eigenspace update, Eq. 6.14
    $\mathbf{Z}^j(k) = \boldsymbol{\Psi}^{j\,T}(k)\mathbf{Y}^j(k)$ ▹ Principle Comp., Eq. 3.21
    $U^j(k) = \frac{1}{m}\sum_{i=1}^{m}\sum_{n=1}^{m} \mathbf{V}^j_{i,n}(k) \times \mathbf{Z}^j_{i,n}(k)$ ▹ Eq. 3.23
    Output: Denoised Response: $\mathbf{U}(k)$
  end for
  → *Beginning of RPCA module :*
  $C(k) = \frac{1}{k}\mathbf{U}(k)\mathbf{U}(k)^T$ ▹ Covariance Estimate
  $C(k) = [\mathbf{W}_k][k\boldsymbol{\Theta}_k][\mathbf{W}_k]^T$ ▹ EVD of Covariance, Eq. 3.24
  $\mathbf{W}_k = \mathbf{W}_{k-1}\boldsymbol{\Lambda}_k \;\&\; \boldsymbol{\Upsilon}_k = \frac{\mathbf{D}_k}{k}$ ▹ Eigenspace Update, Eq. 5.11
  $Z(k) = \mathbf{W}(k) \times \mathbf{U}(k)$ ▹ Transformed Response
  $D^T_M(k), D^R_M(k)$ ▹ Compute R.M.D, ([7])
  $a_1, a_2 : \vartheta(k) = a_1(k)\vartheta(k-1) + a_2(k)\vartheta(k-2)$ ▹ Compute TVAR, ([8])
end for

Ensure: Damage Identification

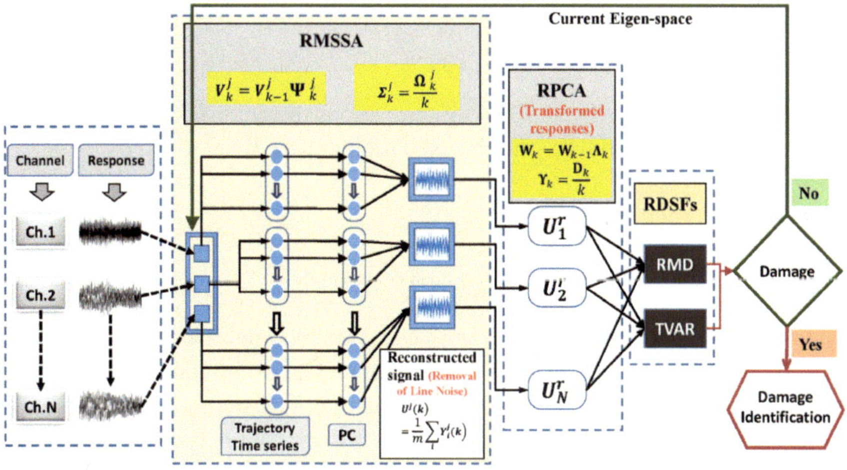

Figure 3.1: Basic framework of proposed RMSSA-RPCA algorithm for real time filtering and damage detection in the presence of operational conditions.

3.5 Numerical models

To demonstrate the applicability of the proposed algorithm for linear and non-linear systems, numerical case studies on structural systems are carried out. In this context, (i) a 5-DOF linear structural system and (ii) a 5-DOF base-isolated systems modeled with Bouc-Wen oscillator at its base with sudden damage in the story stiffness are undertaken. The sudden damage in the systems is induced through a reduction in the story stiffness. The motive behind the introduction of sudden damage through the change in the stiffness is acquired from the theoretical background established in the literature [4–6]. Numerical simulation of the aforementioned structures and their damage detection results are discussed in the present section.

3.5.1 5-DOF linear structural system

A 5-DOF linear vibrating structure modeled as mass, spring and dash-pot system with sudden local damage in the 3rd story is taken for numerical simulation. The input excitation is modeled as *sinusoidal signal* (harmonic) embedded in *White Gaussian noise* (**WGN**). The harmonic component with a cyclic frequency of $15Hz$ accounts for the line noise present in the signal and the increase in energy of line noise is controlled through a decrement in the level of signal to noise ratio (**SNR**). The equation of motion of the 5-DOF linear system can be expressed as Eq. 5.1. The state-space representation for the system subjected to a force vector $F(t)$ can be written as,

$$\begin{aligned} \dot{U} &= \mathbf{A}U + \mathbf{B}F \\ Y &= \mathbf{C}U \end{aligned} \tag{3.26}$$

where, U = vector of states and Y = system response vector governed by the $\mathbf{C}$ matrix. The system matrix, $\mathbf{A}$, and the excitation matrix $\mathbf{B}$ are given by,

$$\begin{aligned} \mathbf{A} &= \begin{bmatrix} [\mathbf{0}]_{5\times5} & [\mathbf{I}]_{5\times5} \\ -\mathbf{M}^{-1}\mathbf{K} & -\mathbf{M}^{-1}\mathbf{C} \end{bmatrix} \\ \mathbf{B} &= \begin{bmatrix} 0 & 0 & 0 & 0 & 0 & -\frac{1}{m} & -\frac{1}{m} & -\frac{1}{m} & -\frac{1}{m} & -\frac{1}{m} \end{bmatrix}^T \end{aligned} \tag{3.27}$$

The mass of the system is considered as $10kg$ at each floor level and the stiffness of the individual floors are assumed to be $7kN/m$ and natural frequencies obtained are $1.2Hz, 3.5Hz, 5.5Hz, 7.1Hz, 8.1Hz$. The damping ratio for the system is kept at $\zeta = 2.0\%$ critical for each mode. The system excitation is simulated using Gaussian white noise. In order to test the sensitivity of the proposed algorithm the random excitation is corrupted with sinusoidal signal which act as line noise in this context. The excitation is simulated for a total duration of $90s$ at a sampling frequency of $100Hz$. Local damage in this system is induced by reduction in the stiffness of 3rd story. Damage cases of 25%, 20% and 15% are taken for the present study. The signal-to-noise ratios (SNR) are considered as 1 and 20 for each investigation.

3.5.2 5-DOF Base-isolated structural system

A non-linear hysteric 5-DOF system modeled as Bouc-Wen oscillator stationed at base with damage in the base story is undertaken in the present section to study the effectiveness of the proposed algorithm in the presence of operational condition. The base equation of motion can be represented as,

$$\mathbf{MX}(t) + \mathbf{CX}(t) + \mathbf{KX}(t) = \mathbf{F}(t) - LS \tag{3.28}$$

Where, **M**,**C** and **K** have their usual notation. The vector **X**(t) represents the displacement response at each floor. **F**(**t**) is input for excitation and has characteristic of *White Guassian Noise*. $L = [1,0,0,0,0]^T$ identifies the location of Bouc-Wen oscillator and the non-linear hysteresis force is presented by S that could be represented as, $S = k_b \lambda X_b(t) + \kappa k_b Z(t)(1-\lambda)$, where, k_b and X_b are Bouc-Wen stiffness and displacement respectively. The non-linear force S is comprised of linear elastic force: $k_b \lambda X_b(t)$ and non-linear hysteric force $k_b Z(t)(1-\lambda)$, where λ is the measure of participation of linear force acting at the base and the remaining part is exerted by hysterics force. Z is a evolutionary parameter referred as the hysteresis displacement which can be expressed by the given form:

$$\dot{Z} = -\gamma \left|\dot{X}_b\right| Z|Z|^{\bar{n}-1} - \beta \dot{X}_b |Z|^{\bar{n}} + \hat{A}\dot{X}_b \tag{3.29}$$

The parameters γ, β, $\bar{n}$ and $\hat{A}$ controls the hysteresis behavior of the isolator and κ controls the amount of non-linearity propagation in the system. In the presence of stochastic excitation the present problem constitutes a coupled non-linear structural model that give rises stochastic differential equations (SDEs). Towards obtaining the response trajectories characterized by the SDEs this 5-DOF base isolation system can be solved by Taylor 1.5 strong technique (based on Ito-Taylor expansion). The state space formulation for the system is given by, [7]

$$\begin{array}{c|c|c|c|c} \dot{X}_b(t) = \dot{Y}_1(t) & \dot{X}_2(t) = \dot{Y}_3(t) & \dot{X}_3(t) = \dot{Y}_5(t) & \dot{X}_4(t) = \dot{Y}_7(t) & \dot{X}_5(t) = \dot{Y}_9(t) \\ \ddot{X}_b(t) = \dot{Y}_2(t) & \ddot{X}_2(t) = \dot{Y}_4(t) & \ddot{X}_3(t) = \dot{Y}_6(t) & \ddot{X}_4(t) = \dot{Y}_8(t) & \ddot{X}_5(t) = \dot{Y}_{10}(t) \end{array} \tag{3.30}$$

However, the Taylor 1.5 techniques also facilitates the formulation of evolutionary parameter Z through SDE which requires another state: $\dot{Z}(t) = Y_{11}(t)$. The Eq. 3.28 can be expressed in the form of Ito-diffusion equation as,

$$d\mathbf{Y} = \mathbf{a}(t, X(t)) \times Y + \mathbf{b}(t, X(t)) \times d\mathbf{W} \tag{3.31}$$

where $Y = [Y_1(t)\ Y_2(t)\ Y_3(t)\ Y_4(t)\ Y_5(t)\ Y_6(t)\ Y_7(t)\ Y_8(t)\ Y_9(t)\ Y_{10}(t)\ Z(t)]^T$ and $a(t,X(t))$ and $b(t,X(t))$ are referred as *drift* and *dispersion* matrices [7]. For the formulation of Taylor 1.5 strong certain moments are in the form of $\Im^o(.)$ and $\Im^j(.)$ are need to be evaluated on the drift and dispersion terms. These moments can be expressed in the matrix form as,

$$\begin{aligned} \Im^o g_l(t,\mathbf{X}_t) &= \left(\frac{\partial}{\partial t} + \sum_{i=1}^{m} a_i(t,\mathbf{X}_t)\frac{\partial}{\partial y_i} + \sum_{i=1}^{m}\sum_{j=1}^{m}\sum_{k=1}^{n} \sigma_{ik}\sigma_{jk}\frac{\partial^2}{\partial y_i \partial y_j} \right) g_l(t,\mathbf{X}_t) \\ \Im^j g_l(t,\mathbf{X}_t) &= \sum_{i=1}^{m}\sum_{j=1}^{n} \sigma_{ij}\frac{\partial g_l(t,\mathbf{X}_t)}{\partial y_i} \end{aligned} \tag{3.32}$$

where, m is the number of first order SDE and n is the number of stochastic process. In this system the m and n can be identified as 5 and 5, respectively. $g_l(t,\mathbf{X}_t$ is any non negative function and the index l represents the state of the system. The Final Taylor 1.5 strong formulation of the system can be carried out by arranging and substituting the moments corresponding to each of the states in the following equation [7]. Foe the complete details of the derivation, the readers are referred elsewhere [7, 8].

$$\begin{aligned} Y_{n+1}(t) = \; & Y_n(t) + \mathbf{a}(t,X(t))\Delta t + \mathbf{b}(t,X(t)) + \Im^1\mathbf{b}(t,X(t))\left\{\frac{(\Delta W)^2 - \Delta t}{2}\right\} + \Im^1\mathbf{a}(t,X(t))\Delta Z + \\ & \Im^0\mathbf{a}(t,X(t))\frac{(\Delta t)^2}{2} + \Im^0\mathbf{b}(t,X(t))(\Delta W\Delta t - \Delta Z) + \Im^1\Im^1\mathbf{b}(t,X(t))\left\{\frac{(\Delta W)^3}{6} - \frac{\Delta W\Delta t}{2}\right\} \end{aligned}$$

$$\begin{bmatrix} \Delta W_n \\ \Delta Z_n \end{bmatrix} = \Delta t^{\frac{1}{2}} \begin{pmatrix} 1 & 0 \\ \frac{\Delta t}{2} & \frac{\Delta t}{2\sqrt{3}} \end{pmatrix} \begin{Bmatrix} U_{n1} \\ U_{n2} \end{Bmatrix} \tag{3.33}$$

with U_{n1} and U_{n2} being the Gaussian random variables with mean zero and unit standard deviation. The mass of the system is considered as $10kg$ at each floor level and the stiffness of the individual floors are assumed to be $5kN/m$. The damping ratio for the system is kept at $\zeta = 2.0\%$ critical for each mode. The parameters that controls the hysteresis behavior of the isolator are considered to be as follows: $\gamma = 0.05$, $\beta = 0.05$, $\bar{n} = 3$ and $\hat{A} = 1$. The measure of participation of linear force acting at base is considered as $\lambda = 0.05$. Similar to previous numerical case study the system is simulated using zero mean Gaussian white noise corrupted by sinusoidal noise measured by the SNRs 1 and 20, applied at each floor level. The simulation is carried out for a duration of $90s$ at a sampling frequency $100Hz$. Global damage in this system is simulated by sudden reduction of the parameter κ, representing the change in non-linearity [4–7]. Damage cases of 25%, 20% and 15% are taken for the present study. The signal-to-noise ratios (SNR) are considered as 1 and 20 for each investigation.

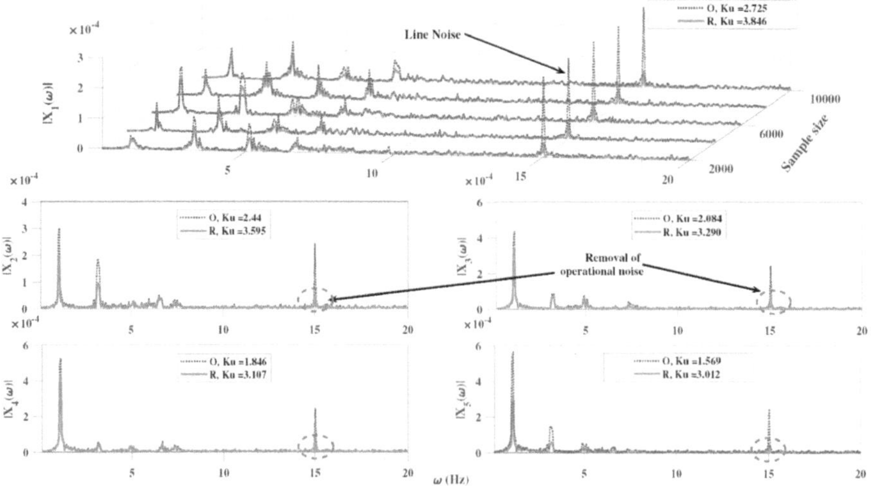

Figure 3.2: Reconstructed signal for Linear 5-DOF system (O-Original, R-Reconstructed, Ku-Kurtosis).

3.6 Damage identification results

3.6.1 RSSA algorithm results: 5-DOF BW-system

Results for 15% change in nonlinearity are provided in this section. Since the transformed responses obtained using the RSSA algorithm consists of simpler components, a relatively low model order (considered here as 2) can be used. The damage instant detected using a combination of AR coefficients and ERD for 15% nonlinearity change at $31s$, is shown in Fig. 3.3. While a change in the mean level of the AR plot verifies the damage instant to be at $31s$, a distinct peak in the recursive ERD plots indicate the exact instant of damage.

3.6.2 5-DOF linear system for RMSSA-RPCA algorithm

To show the applicability of the RMSSA-RPCA algorithm towards the real time filtering and simultaneous damage detection in recursive framework an investigation using the linear 5-DOF system is undertaken. The algorithm processes the system responses (acceleration that is corrupted by sinusoidal signals) as soon as sample points are provided as input. The execution of damage detection takes place through separation of line noise and identification of distortion in sub-space, in real time.

In the present study the harmonic components present in the signal masks the actual information regarding the damage in structure which necessitates the

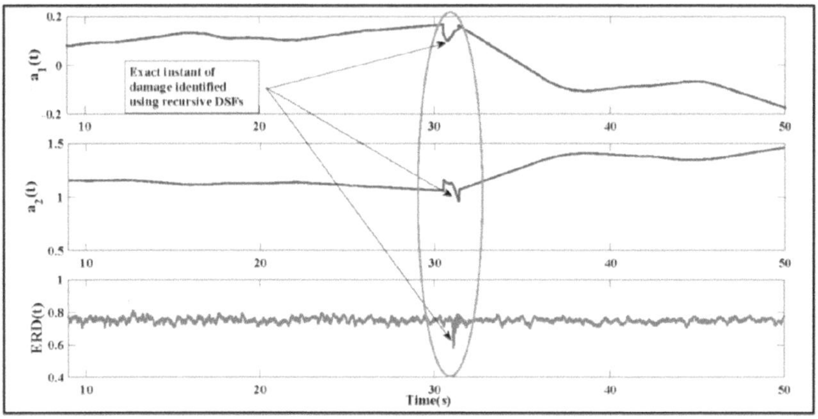

Figure 3.3: RSSA results for 15% global damage.

removal of line noise. The robustness of the RMSSA-RPCA algorithm in identification and significant removal of the noise components from signal can be observed in Fig. 3.2. The figures shows the Fast Fourier Transform (FFT) of the acceleration response corresponding to linear system before and after removal of harmonic component for selected sample points. For clear interpretation between original FFTs (before removal of harmonic component) and reconstructed FFTs (after removal of harmonic component), corresponding *Kurtosis* (*Ku*) indicating the level of harmonic component in the signal is presented along with the FFTs, exploiting the fact that for sinusoidal signals $Ku \leq 3$ and for noisy signals $Ku \geq 3$.

It can be clearly observed from the *original* FFT plots that apart from the five corresponding natural frequencies of each DOFs, there is another peak approximately at 15Hz which represents the presence of line noise in the signal. However when processed with the algorithm it significantly filters out the line component from signal as observed in the *Reconstructed* FFT plot. Thus, it can be clearly observed that the algorithm can remove the signal noise effectively by only leaving small traces of it which proves the efficacy of the RMSSA-RPCA algorithm, towards real time filtering.

For damage detection study the simulated 5-DOF linear system with the local damage in the 3rd story at 50s is undertaken by considering the sudden reduction of stiffness is a representative of local damage in system. The study is carried out for local damage cases of 25%, 20% with 15% reduction in linear story stiffness of the 3rd floor for SNR levels 1 and 20. The response of the linear system used as input to the RMSSA-RPCA algorithm which provides the transformed response devoid of line noise at each time instant. The transformed response is then provided through RDSFs: TVAR and RMD to identify damage instant.

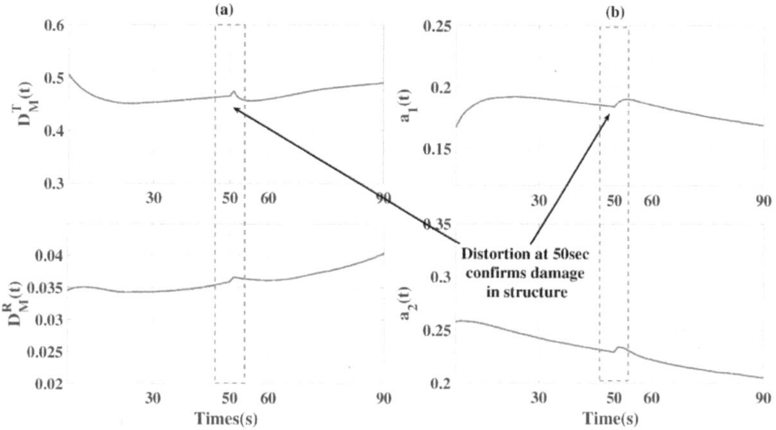

Figure 3.4: Damage detection for 5-DOF Linear system for **SNR 1** with 15% local damage using RMSSA-RPCA.

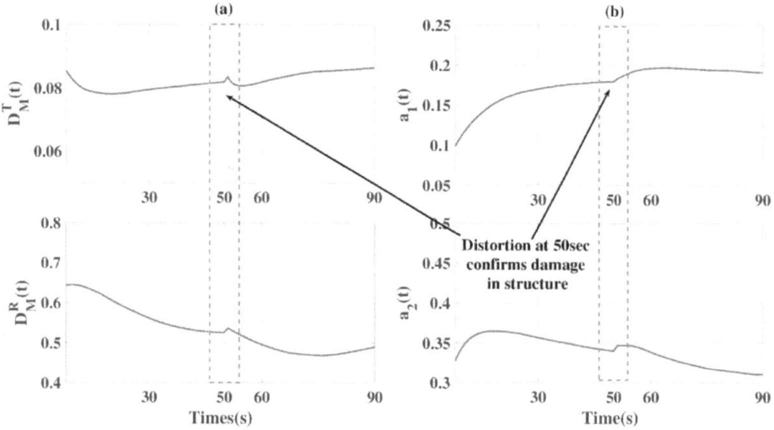

Figure 3.5: Damage detection for 5-DOF Linear system for **SNR 20** with 15% local damage using RMSSA-RPCA.

The damage detection results using TVAR and RMD for SNR 1 with 15% local damage at 50s is portrayed in Fig. 3.4 which exhibits a sharp distortion at 50s which signifies sudden loss in story stiffness in the structure. Similarly for SNR level 20 and 15% reduction at 50s in linear story stiffness the results are displayed in Eq. Figure 3.5 that furnishes the damage detection results using RMSSA-RPCA and their identification through TVAR and RMD for SNR 20 with 15% reduction in stiffness at 50s. The sharp distortion at 50s in Fig. 3.4 and Fig. 3.5 pinpoints the instant of sudden damage in the structure. The damage detection results discussed above clearly proves the robustness of the RMSSA-RPCA algorithm for linear systems.

3.6.3 5-DOF Base-isolated system

In the preceding section the applicability of the RMSSA-RPCA algorithm for damage detection of linear systems after effective removal of the operational conditions is discussed. Here the performance of the RMSSA-RPCA algorithm for real time damage detection in non-linear systems in the presence of line noise is investigated using the base isolated hysteric system.

To demonstrate the efficacy of the RMSSA-RPCA algorithm in filtering, a Hilbert Huang analysis [33] is carried out on the responses of the underlying system. For this, the signal is decomposed into a number of intrinsic mode functions (IMFs) using empirical mode decomposition (EMD). Figure 3.6, which represents FFT of the IMFs before filtering, indicates the presence of operational noise in IMF-1 and IMF-2. After decomposing the signal into IMFs, a spectrogram is plotted in Fig. 3.7 using Hilbert Huang Transform (HHT) for individual IMFs. After acquiring the IMFs, the IMFs corresponding to the operational noise are identified and are discarded to obtain the filtered response.

The damage in the underlying system is modeled as sudden reduction in κ by an amount of 25%, 20% and 15% exactly at 50s. Identification of the instant of damage through TVAR and RMDs are presented in Fig. 3.8 and Fig. 3.9 for 15% reduction in non-linearity with SNR level 1 and 20, respectively.

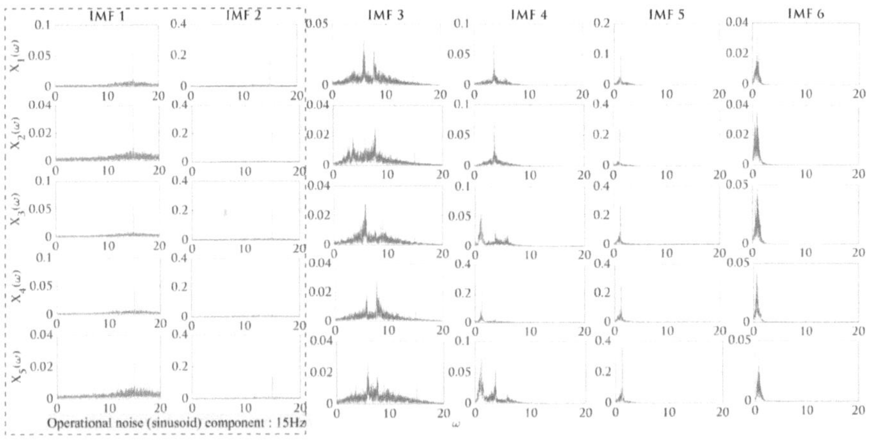

Figure 3.6: FFT of the IMFs of the acceleration responses of 5-DOF base isolated system.

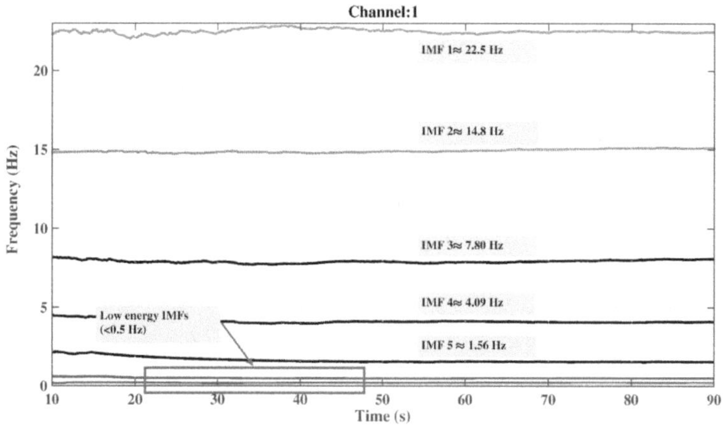

Figure 3.7: Hilbert spectrogram of channel 1 ($X_1(t)$).

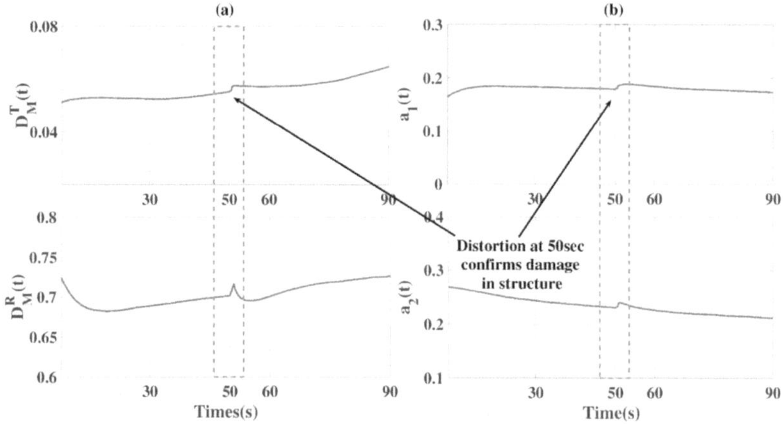

Figure 3.8: Damage detection using TVAR for 5-DOF Base isolated system for **SNR 1** with 15% global damage using RMSSA-RPCA.

Figure 3.8(a) and Fig. 3.9(a) demonstrates the damage detection results using TVAR through a distortion exactly at 50s for SNR 1 and 20 with 15% global damage in the system, respectively. Similarly, both the RMDs exhibit sharp distortion at 50s which indicate the damage due to sudden reduction in base story stiffness. Figure 3.8(b) and Fig. 3.9(b) shows the performance of TVAR towards detection of damage in the system. Sudden jump in mean level of the TVAR coefficients exactly at 50s corroborated bt the results of RMD provides the evidence of the efficacy of the RMSSA-RPCA algorithm in identification of damage in presence of operational conditions even for non-linear systems.

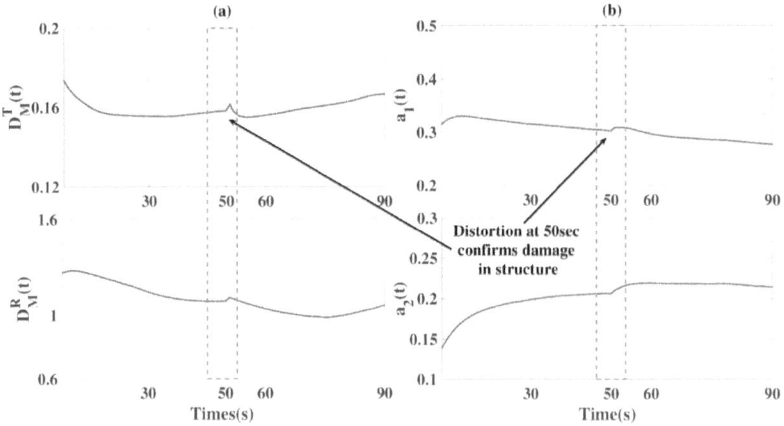

Figure 3.9: Damage detection for 5-DOF Base isolated system for **SNR 20** with 15% global damage using RMSSA-RPCA.

3.6.4 *Threshold for damage detectability of RMSSA-RPCA algorithm*

One of the key aspects of any damage detection algorithm is to understand its limit of detectability. Towards this, an analysis involving the computation of correlation coefficients (ρ) is carried out with different percentages of damage for varied SNR levels in linear and non-linear systems to set a threshold for detectability of damage [5–7]. In linear system, damage is induced locally by reducing stiffness of 3rd floor and for non-linear case change in non-linearity is a cause for global damage. The different percentages of damages considered exactly at 50s are: 25%, 20%, 15% and 12%.

From the online streaming transformed responses obtained using the RMSSA-RPCA algorithm, samples of data are taken where each sample window has the same length. For our study, the windowing is done at an interval of 10s after leaving out few seconds of initial sample. With the construction of a new window (i.e., window-2), the correlation coefficient (ρ) between the new window and the previous one (window-1) is computed. If value of ρ between the windows (ρ_{12}) is significantly low (comparative to other windows) or, negative, then it is further proceeded to compute the value of ρ_{23} to check for any aforementioned traits. If the consecutive values of ρ are negative, or, have low values or, both, then the damage can be interpreted to have occurred in the common window (for example if ρ_{12} and ρ_{23} are significantly low or negative then damage has occurred in window-2). But if window-2 and window-1 has strong correlation then ρ between the successive windows are evaluated until there is any change in sign, or, abrupt decrement in the value of ρ.

Table 3.2: Threshold for damage detection using RMSSA-RPCA algorithm for linear 5-DOF systems.

% of damage	SNR	Correlation coefficients ρ					
		ρ_{12}	ρ_{23}	ρ_{34}	Damage occurrence	ρ_{45}	ρ_{56}
25%	1	0.9650	0.9985	**-0.7712**		**-0.7832**	0.9970
	20	0.9797	0.9966	**-0.6402**		**-0.6557**	0.9980
20%	1	0.8311	0.9980	**-0.6901**		**-0.7276**	0.9960
	20	0.9919	0.9989	**-0.6160**		**-0.6223**	0.9962
15%	1	0.9833	0.9983	**-0.5829**		**-0.5635**	0.9864
	20	0.9247	0.9980	**-0.3849**		**-0.4029**	0.9861
12%	1	0.9312	0.8150	0.6535		0.8991	0.8661
	20	0.9123	0.8603	0.7484		0.6508	0.6502

ρ: Window-1: 18s-28s,Window-2: 28s-38s,Window-3: 38s-48s,Window-4: 48s-58s, Window-5:58s-68s, Window-6: 68s-78s. ρ_{ij} = Correlation coefficient of Window-i&j
The values in bold fonts indicate the detectable range of RMSSA-RPCA algorithm.

Table 3.3: Threshold for damage detection using RMSSA-RPCA algorithm for Non-linear 5-DOF systems.

% of damage	SNR	Correlation coefficients ρ					
		ρ_{12}	ρ_{23}	ρ_{34}	Damage occurrence	ρ_{45}	ρ_{56}
25%	1	0.9825	0.9978	**-0.5405**		**-0.5087**	0.9951
	20	0.9967	0.9937	**-0.4656**		**-0.4619**	0.9778
20%	1	0.9727	0.7589	**-0.5187**		**-0.3606**	0.7044
	20	0.9892	0.9983	**0.1281**		**-0.0149**	0.9910
15%	1	0.8268	0.9178	**0.0408**		**0.1479**	0.7280
	20	0.9983	0.9971	**0.0744**		**0.0141**	0.9936
12%	1	0.9843	0.9580	0.9299		-0.8639	0.8641
	20	0.9107	0.7985	0.5004		0.5163	0.6941

ρ: Window-1: 18s-28s,Window-2: 28s-38s,Window-3: 38s-48s,Window-4: 48s-58s, Window-5:58s-68s, Window-6: 68s-78s. ρ_{ij} = Correlation coefficient of Window-i&j
The values in bold fonts indicate the detectable range of RMSSA-RPCA algorithm.

Table 3.2 and Table 3.3 presents the correlation coefficients between the windows for linear and non-linear systems respectively. ρ_{12}, ρ_{23} and ρ_{56} shows high value of correlation coefficients (each of them ≥ 0.7) whereas ρ_{34} and ρ_{45} are either negative or has significantly low values (e.g., 0.0408, 0.0744, 0.1479). The occurrence of such a trend in the values of ρ_{ij} can be inferred from the fact that the structure remains in a state of stable equilibrium until the damage instant thereby yielding high positive values of correlation coefficient. The negative or significantly low values of ρ_{34} and ρ_{45} can be attributed to the fact that the common window, window-4 contains the damage instant which is the transition between the stable equilibria before and after the damage instant. It is observed that for 12% damage, the values of ρ do not abide by the conditions mentioned in the above steps whereas for 15% damage and beyond the values of ρ_{34} and ρ_{45} either suffers a drastic reduction as compared to the previous one or a change in sign. This sets the lower limit of detectability, i.e., threshold of the RMSSA-RPCA algorithm as 15%.

3.7 Robustness of the RMSSA-RPCA algorithm

3.7.1 Damage detection in the presence of low frequency-operational noise

Measurements during the present experimental study (presented in later section) using adequate instrumentation reveals that the frequency of noise components are usually well separated from that of structural frequency band. But in situations of inadequate sensors, highly noisy environment or interference of noise with structures having closely spaced modes, the problem of low frequency-operational noise in signal may hamper the degree of identification. Apart from the signal noise another reason for the interference of structural modes with noise frequency is due to the environmental conditions. A study on the damage detection of a structural system using the RMSSA-RPCA method, particularly in the presence of low frequency noise components is of prime interest in the present section. In this context, the 5-DOF linear system as described in the aforementioned section is simulated using white Gaussian noise. The simulation of operational noise is carried out at a frequency rate of 2Hz, such that, the fundamental frequency is close to the noise component. The presence of noise near fundamental frequency corrupts the signal with outliers which masks the damage information. This can be attributed to the fact that the structure is primarily dominated by the fundamental frequency, whereas in the presence of noise affects the participation of relative modes in damage detection. The filtering of the low frequency-operational noise and simultaneous damage detection is presented in the Fig. 3.10 and Fig. 3.11, respectively. From the damage detection study it can concluded that the algorithm can substantially classify the noise components even when the mixing of noise frequency and structural mode occurs.

3.7.2 Case study for the 5-DOF base isolated system using non-stationary excitation

For an understanding on the performance of the algorithm to implementation in field applications, the 5-DOF base isolated system modeled is simulated using a non-stationary excitation in the presence of operational condition. The

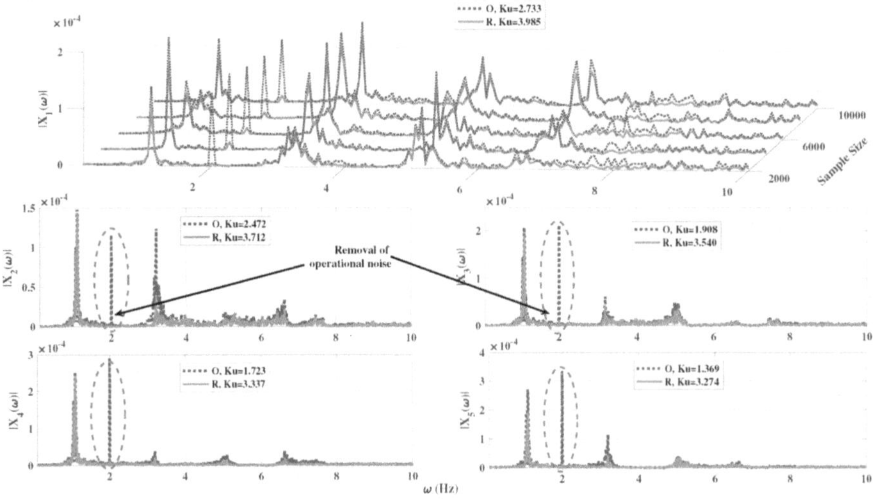

Figure 3.10: Reconstruction of signal with **noise frequency 2Hz**.

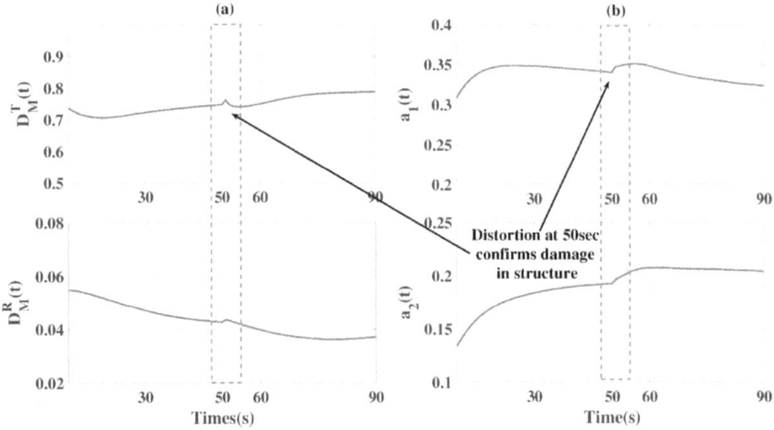

Figure 3.11: Damage detection of 5-DOF linear system for low frequency of operational noise with 15% local damage using RMSSA-RPCA.

non-stationary time series is generated by using Kanai-Tajimi model [34]. The simulation is carried out for a total duration of 50s with bandwidth parameter, standard deviation and dominant frequency as 0.3, 0.4 and 3Hz, respectively. The value of the envelop function is assumed to be 0.3. Global damage in the system is simulated using sudden reduction in the non-linearity (κ) of the system exactly at 30s. Operational noise is introduced in the system acceleration response fol-

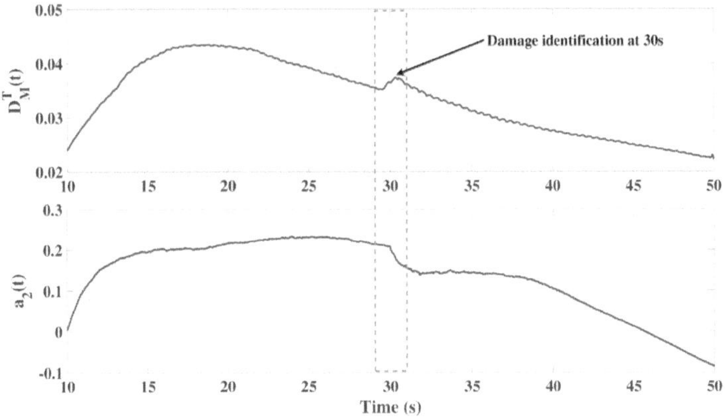

Figure 3.12: Damage detection of 5-DOF Base isolated system for **non-stationary excitation** with 15% global damage using RMSSA-RPCA.

lowing the similar steps as previously stated. For this case study a SNR of 10 is undertaken. From the results presented in Fig. 3.12 it can be inferred that the RMSSA-RPCA algorithm is successful in significant removal of noise effect and identifies the exact instant of damage at 30s.

3.7.3 Comparison with RPCA algorithm

The previous section demonstrated the applicability of RMSSA-RPCA algorithm towards noise sensing and damage detection in a recursive framework for both linear and nonlinear systems. From the aforementioned theoretical background and result discussions it can be summarized that the RMSSA-RPCA algorithm can remove the noise component from the response signals which makes successful damage detection possible. In this section the advantages of RMSSA-RPCA algorithm over recently established recursive identification tool namely recursive principal component analysis (RPCA) [5,6] is illustrated through numerical comparison (Table 3.4). The RPCA algorithm has shown ability to identify the damage instants in real time [5,6]. However, an investigation reveals that in the presence of line noise, RPCA fails to significantly identify the changes in eigenspace, thereby producing erroneous detection. In this regard, the damage detection results of numerical systems simulated against different levels of SNR provides a test-bed for comparing the performance of RMSSA-RPCA with RPCA algorithm in the presence of line noise. The improved performance of the RMSSA-RPCA algorithm over RPCA in the presence of operational noise is exemplified by comparing the damage detection results over linear and non-linear systems and presented in Table 3.4. However, the results are discussed in detail only for the non-linear

Table 3.4: Summary of real time damage detection results.

Type of systems	Damage in %	RDSFs	% Change in Mean Level			
			RMSSA-RPCA		RPCA	
			SNR 20	SNR 1	SNR 20	SNR 1
Linear 5DOF System	25%	TVAR	11.39	12.88	12.08	2.52
		RMD	16.36	18.51	10.53	2.87
	20%	TVAR	27.65	10.90	8.50	-
		RMD	12.28	20.22	12.45	-
	15%	TVAR	12.67	8.35	2.60	-
		RMD	12.73	4.12	0.38	-
Base isolated 5DOF System	25%	TVAR	10.36	8.17	7.47	2.52
		RMD	21.62	19.64	1.04	2.87
	20%	TVAR	10.29	11.91	6.60	-
		RMD	15.18	9.50	10.61	4.98
	15%	TVAR	7.30	10.14	3.02	-
		RMD	9.73	5.29	0.078	-

Damage is simulated through change in story stiffness, k, for linear and change in non linearity, κ, for non-linear system.
The empty cell refers to the inability of respective algorithm in identification of damage instant.

system modeled as Bouc-Wen for brevity. Towards this, three separate cases with different levels of SNR, viz., SNR = 20, 10 and 1 which controls the energy of line noise is considered to show the affect of line noise in damage detection.

Detection results for the non-linear 5-DOF system using RPCA algorithm with 20% reduction in stiffness for varied levels of SNRs is undertaken to investigate the sensitiveness of the algorithm against non-linear systems. The subplots (a),(b) and (c) in Fig. 3.13 corresponds to damage detection against SNR levels 20,10 and 1, respectively. Referring to Fig. 3.13, it can be clearly understood that for SNR=20 the RPCA algorithm can successfully identify the damage instant. But for SNR=10 the detection using RPCA becomes uncertain owing to the increased jaggedness due to increase in the amplitude of the operational noise (modeled as sinusoid). Thus it is clear from the observations that, more the amplitude of the sinusoid, poorer the detectability using RPCA. The RMSSA-RPCA algorithm provides significant improvement even in presence of substantial operational noise. In this context, the RMSSA-RPCA algorithm can clearly distinguish between the line noise and damage characteristics from the signal thereby facilitating successful identification of the damage instances as observed in the Fig. 3.8 and Fig. 3.9. This provides the evidence for advantage of RMSSA-RPCA algorithm which has ability to detect damage irrespective of the energy content of operational noise.

Table 3.4 summarizes the detectability of RMSSA-RPCA and RPCA algorithm in the presence of operational noise through the percentage change in mean

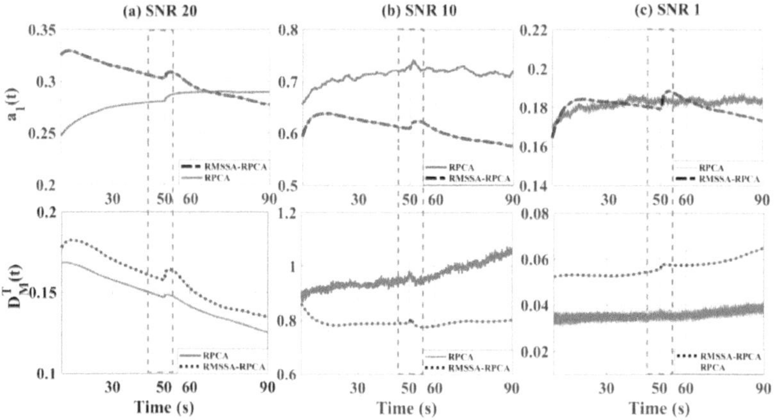

Figure 3.13: Damage detection of Bouc-Wen 5-DOF system using RPCA+TVAR for 20% damage (SNR is interpreted as the ratio of σ of WN to amplitude of operational noise/sinusoid).

level before and after damage. It can be observed from Table 3.4 that RPCA algorithm which neglects the effect of operational condition for damage detection detects the damage instant and shows significant change in mean level for SNR 20 (i.e., low energy of line noise) but detectability is compromised for most of the cases as the SNR level increases to 1 (i.e., high energy of line noise). Whereas the RMSSA-RPCA algorithm which considers the filtered signal for the construction of transformed responses, effectively identifies the damage instant and shows significant change in mean level for both SNR 20 and 1.

3.8 Experimental verification

To verify the robustness of the proposed algorithm towards real time damage detection in the presence of line noise, the proposed algorithm is utilized on a laboratory experiment modeled under controlled environment to separate the effect of line noise and identify the damage in real time.

3.8.1 Description of the setup and measurement program

An aluminum cantilever beam of dimensions $135cm \times 3.5cm \times 0.5cm$, fixed on a base plate which is a combination of two plate sections of dimension $23cm \times 15cm \times 1cm$ welded together to ensure right angle is used for the experimental verification. A rubber sample of stretched length 20cm is attached to the free end of beam which acts as a source of non-linearity to the system [8]. The base plate is fixed on a shake table with bolted connections. The excitation used in the experimental study is modulated white noise (non-stationary) as shown in the Fig. 3.15. The details of excitation data, sensors and shake table is shown in Table 3.5.

The position of the accelerometers on the beam are 42cm, 65cm and 95cm respectively from the free end. A heat source is placed under the rubber sample to simulate the damage. To keep a measure of the rise in temperature a thermocouple module (MAX6675) consisting of 3 numbers of K-type sensor (measure capacity 1024°C) is utilized. The temperature data is utilized to control the preheating temperature so as to allow a sufficient length of time before complete snapping of rubber sample. Preheating is carried out to raise the initial temperature of the rubber sample to $59°C$ to demonstrate a practical case study where in a structure undergoing random vibration in the presence of temperature rise suffers a quick damage. The modeling of degradation is not pursued in present case study as the rubber strip is controlled to fail faster by preheating. Real time damage detection of degrading systems in the presence of operational condition needs to be taken up in future as an extension of the present work. In the course of 88s, the temperature in the rubber sample rose to $94.5°C$ at which the snapping of the rubber sample was observed approximately at 12cm from the free end of the beam. The snap off of the rubber sample corresponds to near sudden damage in the structure. The presence of line noise in experimental data is inevitable as it is generally induced by high voltage power source which drives the shake table and also acts as a power source for the data acquisition system. The acceleration data obtained from the deck is than processed by the proposed algorithm to detect the instant of snapping of the rubber sample by removing the inherent line noise assuming that a new data is recorded at every instant of time. The arrangement of the experimental setup with the positioning of the equipment used is presented in the Fig. 3.14.

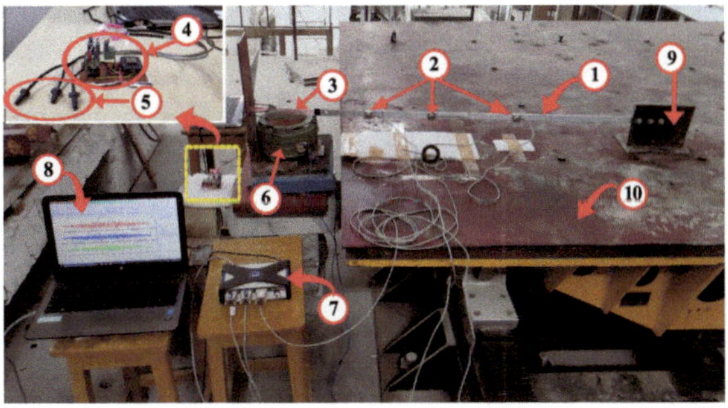

Figure 3.14: Experimental setup of Aluminium Beam Experiment, 1. Aluminium Beam, 2. Placement of Accelerometers, 3. Rubber strip, 4. Thermocouple Module, 5. K-Type Heat sensors, 6. Heater, 7. Data Acquisition system, 8. Data recording, 9. Base plate, 10. Shake table.

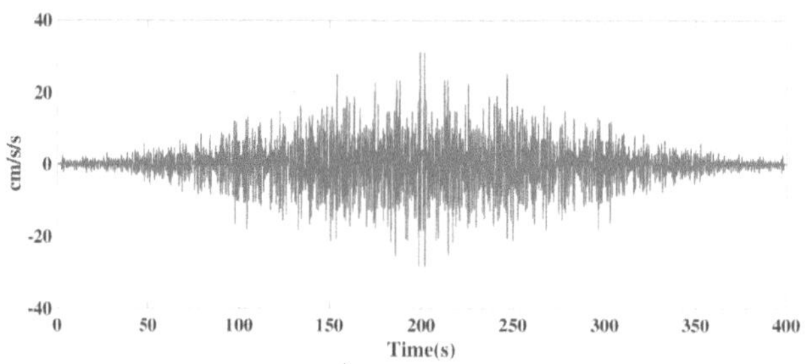

Figure 3.15: Input excitation.

Table 3.5: Summery of excitation data, sensors and shake table used in experimental study.

Data acquisition	
Type of data	Zero mean Gaussian white noise passed through a hamming window
Sampling rate	150 Hz
Duration of excitation	400 s
Data acquisition system	Quantum X MX410 HBM
Type of sensors	Honeywell accelerometers TEDS
frequency range of sensors	0 Hz to 25 kHz
Shake table (Model no. Bi-00-300)	
Dimension	150cm×150cm
Payload capacity	$5ton$
Peak acceleration	$\pm 2g$
Peak Displacement	$\pm 50cm$
Frequency range	0 Hz to 20 Hz

3.8.2 Results

The primary steps in the implementation of RMSSA-RPCA algorithm contains the collection of few initial sample points to form a initial sub-space and recursive update of the sub-space through the use of FOEP technique after separation of operational noise from the acceleration response and finally the detection of damage instances using the RDSFs.

For understanding the dynamical behavior of the setup, whether linear or nonlinear, the system is excited with different amplitude of the modulated white noise. Figure 3.16 shows the FFT plot for modulated white noise with the amplitude 0.75, 0.88 and 1. The shift in the fundamental frequency due to change in amplitude of the forcing function can be depicted in Fig. 3.16, which verifies the non-linearity of the taken setup. The presence of operational noise from the signal is represented in Fig. 3.17 through the Hilbert spectrogram of acceleration responses. From Fig. 3.17 it can be inferred that IMF-1 corresponds to the operational noise. For filtering through RMSSA algorithm the components other than

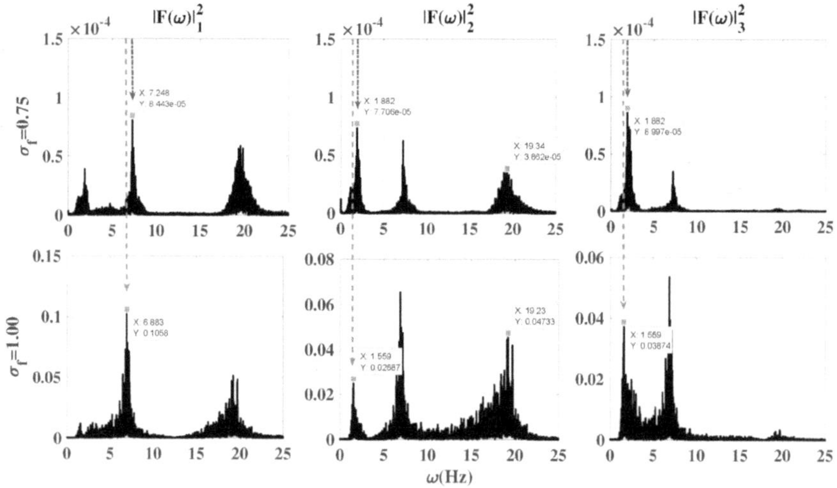

Figure 3.16: Non-linearity through FFT of response scaled to varied levels.

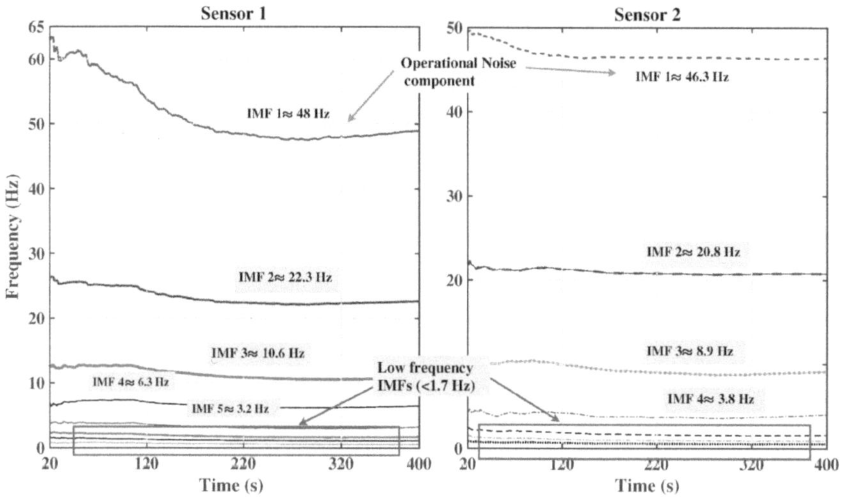

Figure 3.17: Hilbert spectrogram of sensor-1 and sensor-2.

the operational noise (i.e., IMF-1) is taken and the damage analysis is carried out on the filtered response. Figure 3.18 shows the shift in frequency through FFT plot of acceleration response of the beam in healthy and damaged state. It can be visualized that there is a reduction of system frequency thus indicating damage in the beam.

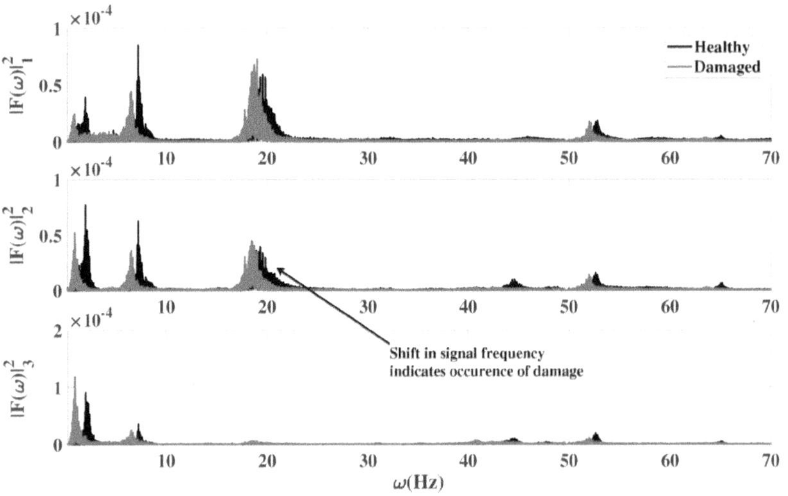

Figure 3.18: Comparison in FFT of response of beam before and after damaged.

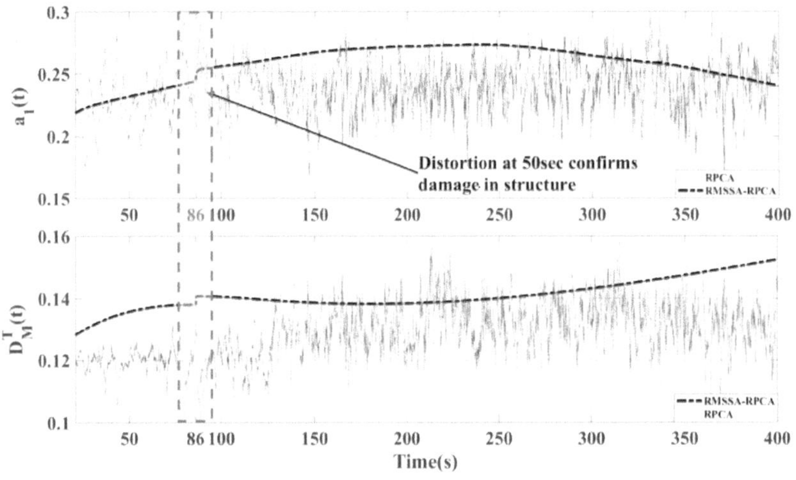

Figure 3.19: DSFs of beam at damaged state.

The damage detection using proposed algorithm in conjunction with RDSFs is then carried out on the RMSSA-filtered responses in real time. The results for damage detection using TVAR and RMD are presented in Fig. 3.19. Furthermore, To illustrate the advantage of the proposed framework over RPCA algorithm, the identification results of RPCA is portrayed in the Fig. 3.19. It is observed from the TVAR coefficients that a sharp distortion occurs approximately at 88s thereby capturing the instant of snapping of rubber sample successfully in real time which corroborates with results of RMD. It can be observed that after the identification of damage instant, the RDSFs follow a trend which imitates that of the input excitation in Fig. 3.15. However RPCA algorithm fails to identify the damage instant owing to the fact that it requires pre-processing of data in order to separate the operational noise for effective damage detection. Hence it can be safely concluded that the proposed RMSSA-RPCA algorithm is robust even in the presence of operational noises and can extract the damage information from the signal to facilitate real time damage detection in structural systems and thereby is applicable to field implementations.

3.9 Conclusions

The chapter demonstrates a robust structural damage detection algorithm under the presence of operational conditions using RMSSA and RPCA in an unified framework. Towards simultaneous filtering and real time damage detection in real time the proposed algorithm utilizes first order perturbation technique for evaluation of recursive updates of eigenspaces in RMSSA and RPCA modules. The recursive updates of eigenspace in RMSSA facilitates removal of noise components in a real time framework and the subspace updates at each successive instant enables damage identification on the filtered response. The detection of damage under operational conditions using the proposed framework has demonstrated successful identification for sudden damages of order 15% for both linear and non-linear systems. However, the for sudden damages of order less than 15% the proposed framework fails to identify the damage instant. Furthermore a case study on the recently developed technique RPCA has illustrated a significant decrease in its performance with the increase in the energy of operational noise, thus clearly depicting the advantage of the proposed framework over RPCA in the presence of operational conditions. Towards field implementation of the proposed algorithm a case study using Kanai-Tajimi type earthquake models has clearly illustrates the efficacy of the proposed framework in damage identification even for non-stationary type of excitation.

The proposed algorithm provides a framework to account for the first order error approximations in recursive estimation of subspace in FOEP technique. Towards this the proposed algorithm assumes an initial estimation of error based on numerical studies. However, for accurate identification of damage instant, the estimation and incorporation of errors in real time is a crucial aspect. This limitation requires simultaneous estimation of LNMs and POMs to accurately incorporate the errors which is beyond the scope of current framework and needs to be carried out as an extension of the present study.

References

[1] Chanpheng, T., Yamada, H., Katsuchi, H. et al. (2012). Nonlinear features for damage detection on large civil structures due to earthquakes. Structural Health Monitoring, 11(4): 482–488.

[2] Neu, E., Janser, F., Khatibi, A. A. et al. (2016). Automated modal parameter-based anomaly detection under varying wind excitation. Structural Health Monitoring, 15(6): 730–749.

[3] Koshimura, S., Namegaya, Y. and Yanagisawa, H. (2009). Tsunami fragility—A new measure to identify tsunami damage. Journal of Disaster Research, 4(6): 479–488.

[4] Bhowmik, B., Krishnan, M., Hazra, B. et al. (2018). Real-time unified single-and multi-channel structural damage detection using recursive singular spectrum analysis. Structural Health Monitoring, 1475921718760483.

[5] Krishnan, M., Bhowmik, B., Tiwari, A. et al. (2017). Online damage detection using recursive principal component analysis and recursive condition indicators. Smart Materials and Structures, 26(8): 085017.

[6] Krishnan, M., Bhowmik, B., Hazra, B. et al. (2018). Real time damage detection using recursive principal components and time varying auto-regressive modeling. Mechanical Systems and Signal Processing, 101: 549–574.

[7] Tripura, T., Bhowmik, B., Pakrashi, V. et al. (2019). Real-time damage detection of degrading systems. Structural Health Monitoring, 1475921719861801.

[8] Bhowmik, B., Tripura, T., Hazra, B. et al. (2019). First-order eigen-perturbation techniques for real-time damage detection of vibrating systems: Theory and applications. Applied Mechanics Reviews, 71(6).

[9] Sadhu, A., Hazraa, B. and Narasimhan, S. (2014). Ambient modal identification of structures equipped with tuned mass dampers using parallel factor blind source separation. Smart Structures and Systems, 13(2): 257–280.

[10] Sadhu, A., Hazra, B. and Narasimhan, S. (2013). Decentralized modal identification of structures using parallel factor decomposition and sparse blind source separation. Mechanical Systems and Signal Processing, 41(1-2): 396–419.

[11] Chang, J. C. and Soong, T. T. (1980). Structural control using active tuned mass dampers. Journal of the Engineering Mechanics Division, 106(6): 1091–1098.

[12] Sohn, H. (2006). Effects of environmental and operational variability on structural health monitoring. Philosophical Transactions of the Royal Society A: Mathematical, Physical and Engineering Sciences, 365(1851): 539–560.

[13] Loh, C. H. and Chen, M. C. Modeling of environmental effects for vibration-based shm using recursive stochastic subspace identification analysis. In Key Engineering Materials, volume 558. Trans. Tech. Publ., pp. 52–64.

[14] Yan, A. M., Kerschen, G., De Boe, P. et al. (2005). Structural damage diagnosis under varying environmental conditions—part I: A linear analysis. Mechanical Systems and Signal Processing, 19(4): 847–864.

[15] Adams, D. (2007). Health Monitoring of Structural Materials and Components: Methods with Applications. 1 ed. John Wiley & Sons.

[16] Cahill, P., Hazra, B., Karoumi, R. et al. (2018). Vibration energy harvesting based monitoring of an operational bridge undergoing forced vibration and train passage. Mechanical Systems and Signal Processing, 106: 265–283.

[17] Kerschen, G., Poncelet, F. and Golinval, J. C. (2007). Physical interpretation of independent component analysis in structural dynamics. Mechanical Systems and Signal Processing, 21(4): 1561–1575.

[18] Jolliffe, I. (2010). Principal Component Analysis. 2nd ed. Springer Series in Statistics, Springer.

[19] Broomhead, D. S. and King, G. P. (1986). Extracting qualitative dynamics from experimental data. Physica D: Nonlinear Phenomena, 20(2-3): 217–236.

[20] Loh, C. H., Chen, C. H. and Mao, C. H. (2010). Detecting seismic response signals using singular spectrum analysis. Sensors and Smart Structures Technologies for Civil, Mechanical, and Aerospace Systems, volume 7647.

[21] Rousseeuw, P. J. and Leroy, A. M. (1987). Robust Regression and Outlier Detection, volume 1. Wiley Online Library.

[22] Jouan-Rimbaud, D., Massart, D., Saby, C. et al. (1998). Determination of the representativity between two multidimensional data sets by a comparison of their structure. Chemometrics and Intelligent Laboratory Systems, 40(2): 129–144.

[23] Vandeginste, B., Massart, D., De Jong, S. et al. (1998). Handbook of Chemometrics and Qualimetrics: Part B. Elsevier.

[24] Wu, W., Mallet, Y., Walczak, B. et al. (1996). Comparison of regularized discriminant analysis linear discriminant analysis and quadratic discriminant analysis applied to nir data. Analytica Chimica Acta, 329(3): 257–265.

[25] Derde, M. P. and Massart, D. (1986). Uneq: A disjoint modelling technique for pattern recognition based on normal distribution. Analytica Chimica Acta, 184: 33–51.

[26] Golyandina, N. and Zhigljavsky, A. (2013). Singular Spectrum Analysis for Time Series. 1 ed. Springer Briefs in Statistics, Springer-Verlag Berlin Heidelberg.

[27] Gruszczynska, M., Rosat, S., Klos, A. et al. (2019). Multichannel singular spectrum analysis in the estimates of common environmental effects affecting gps observations. pp. 211–228. In Geodynamics and Earth Tides Observations from Global to Micro Scale. Springer.

[28] Rangelova, E., Sideris, M. and Kim, J. (2012). On the capabilities of the multi-channel singular spectrum method for extracting the main periodic and non-periodic variability from weekly grace data. Journal of Geodynamics, 54: 64–78.

[29] Mirmomeni, M., Lucas, C., Araabi, B. et al. (2011). Recursive spectral analysis of natural time series based on eigenvector matrix perturbation for online applications. IET Signal Processing, 5(6): 515–526.

[30] Kreyszig, E. (1999). Advanced Engineering Mathematics, 8th edition.

[31] Feeny, B. (2002). On proper orthogonal co-ordinates as indicators of modal activity. Journal of Sound and Vibration, 255(5): 805–818.

[32] Sadhu, A. and Hazra, B. (2013). A novel damage detection algorithm using time-series analysis-based blind source separation. Shock and Vibration, 20(3): 423–438.

[33] Peng, Z., Peter, W. T. and Chu, F. (2005). An improved hilbert–huang transform and its application in vibration signal analysis. Journal of Sound and Vibration, 286(1-2): 187–205.

[34] Lin, Y. and Yong, Y. (1987). Evolutionary kanai-tajimi earthquake models. Journal of Engineering Mechanics, 113(8): 1119–1137.

Chapter 4

Multi-sensor Real Time Damage Detection Techniques (A): RPCA

The preceding chapters have focussed on the principles of structural damage detection approaches and their mathematical preliminaries. At this point, one is now faced with the challenge of making an accurate assessment of an operating condition of a structure based on the vibration response extracted at a particular time stamp. Once *relationships* from the data are derived, the map between features and diagnosis can be constructed in the backdrop of multi-sensing strategies employing recursive principal component analysis (RPCA) as an automated SHM process. The use of RPCA can be treated as the sophisticated extension of PCA explored in a real-time format. Towards this, the chapter first discusses the Data to Decision approach based on the discipline of machine learning – more specifically, its *pattern recognition* aspects – to learn relationships from data. The use of pattern recognition allows the possibility of automating SHM. Embedded in its details is a general structural dynamical system cast in the framework of RPCA. The formulation for RPCA is presented next, followed by the results of numerical simulation on a simple Buoc-Wen (B-W) model. A separate case study providing identification results for an under determined case is also explored. Finally, a comparison between traditional PCA and its automated variation – the RPCA, is provided – that highlights the superiority of recursive estimates towards exact identification of damage instants and location in a unified framework. For

Table 4.1: Important acronyms.

PCA	Principal Component Analysis
RPCA	Recursive Principal Component Analysis
FOEP	First Order Eigen Perturbation
EVD	Eigen Value Decomposition
PR	Pattern Recognition
RRE	Recursive Residual Error
EVC	Eigen Vector Change
CI	Condition Indicator
B-W	Buoc Wen
MDOF	Multi-Degree of Freedom

easy comprehension of the subsequent discussions, some useful acronyms are provided in Table 4.1.

4.1 Motivation

For physical infrastructures, a global NDE technique utilizing the vibration characteristics of a monitored system is effective in assessing the condition of the overall structure. Regardless of the choice of algorithms, an important step in vibration-based SHM is to extract meaningful features from output measurements. The most common features of interest used to detect structural damage patterns are changes in the eigenstructure of the covariance estimates – leading to deviations in natural frequencies and mode shapes [7, 9, 23, 24]. In cases where excitation forces cannot be measured, adequate modal analysis techniques can be used on the sensor outputs to extract these features from ambient vibrations [10]. As compared to prevalent input-output excitation techniques, the measurement of the structural response using ambient excitation is performed while the structure is in service (i.e., no evicting building residents, no downtime maintenance for turbines, no bridge closures). The technique is cost effective (essentially no cost for the excitation source) and can be abundantly available to monitor physical infrastructure. Table 4.2 provides a quick summary of the merits associated with using these methods. Progress – especially in the last few decades – in sensors and instrumentation have led to installation of dense array network models for large infrastructures such as bridges, that aim at monitoring the system health in real-time. The difficulty in selecting appropriate algorithms for damage detection lie in their formulation – which are mostly not suited to analyse such a large amount of data in real-time – that can be alleviated by the use of recursive FOEP methods discussed in detail in the subsequent sections. Moreover, most analysis techniques require the need for human intervention and user interactions, although some efforts have been made to automate this process [6].

Table 4.2: Performance of output-only techniques.

Output-only	Advantages	Disadvantages
Ambient input excitation	Cost-effective (free excitation source)	Uncontrolled amplitude of excitation input to the structure
	Widely applicable - bridges, buildings, wind turbines	Difficult to excite lateral and higher torsional modes of bridges and tall structures
	Can be adopted for real-time continuous monitoring	Low frequency excitation (< 1 kHz)

The use of data driven techniques like PCA have gained significant popularity in SHM literature in recent times [1–5]. PCA, a technique for extracting eigenspace characteristics using response only data, can be used as an indicator of the system's existing state – which in turn can be used to detect changes in a structure's health over time. However, PCA frequently requires the use of a reference (*baseline*) data and operates mainly in an offline batch mode that impedes its applicability towards online SHM implementation. Moreover, for cases where cost impediments are predominant, PCA cannot be used to detect damage patterns using a single sensor and requires special modifications in its formulation [17]. The aforementioned practical problems motivate the need for a method that extends the confines of PCA in real-time to detect damage for both linear and nonlinear systems. This can be accomplished by recasting the problem of damage detection in a perturbative framework, that provides real-time updates of the eigenstructure. This method is termed as recursive principal component analysis (RPCA) [8, 10].

4.2 Problem formulation

The use of PCA in damage detection is manifested by the use of windowed data samples that are acquired over a fixed period of time. The approach to real-time SHM discussed here is based on the idea of pattern recognition (PR). PCA, an inherently baseline-reliant algorithm, fails to perform in an online framework due to the choice of parameters (i.e., window size, overlap, shift and more) that are mostly arbitrary and problem specific. In order to make PCA amenable towards real-time damage detection, modifications in its formulation is hereby proposed using the FOEP approach discussed in the preceding chapters. That is, the RPCA based framework is associated with the rank-one update of the eigenspace of the covariance matrix applied to the data vector X_k [12]. In the broadest sense, spectral decomposition of the response covariance matrix yields eigenvalues and eigenvectors, that correspond to natural frequencies and modeshapes, respectively. Tracking the eigenstructure changes in real-time provides identification of possible damage. One primary advantage of the RPCA approach is the fact that damage can be detected without the requirement of data from every conceivable damage situation – thereby reducing serious modelling demands necessary for supervised learning. Unlike most PR algorithms that require a training diagnostic to produce a correct class label, the proposed recursive formulations only compare the damage pattern from the previous time stamp and infer the possibility of a fault [8–10]. The other advantage of this approach lies in the significant reduction in computational complexity, which is reflected from the recursive updates of eigenstructure – contrary to updating the entire covariance matrix at each iteration – thereby providing immediate eigenspace updates without the added complexity [13].

4.3 Recursive covariance estimation and FOEP

The practice of employing PCA for vibration based health monitoring is well documented in the literature [15, 18]. In order to tailor basic PCA towards online damage detection, algorithms premised on FOEP are chosen that update the eigenspace at each time step [8, 10]. In order to explain this, a digression on the terminology of pattern recognition is essential. It is imperative to understand that the basic concepts of FOEP are embedded as a data driven approach. It should be noted that as the method is purely data driven, it does not make any *apriori* assumptions regarding the nature of the data – thereby finding applications for binary data sets, data evolving from chemical engineering processes [19] and even in structural dynamics problems. This mode of unsupervised learning almost exclusively applies to fault detection and localization diagnostics – commonly referred to as *novelty detection* or *anomaly detection* methods [20–22] – where statistical features of interest are investigated to infer damage patterns of the monitored system.

FOEP is a way of expressing the eigen structure of the $k+1^{th}$ step in terms of the eigen structure of the kth step as the $k+1^{th}$ data streams in. This is accomplished by expressing the eigenvalue decomposition (EVD) of symmetric positive definite covariance matrix in terms of rank one perturbation of eigenvalue and eigenvector matrices [8,10]. Consider the data matrix $\mathbf{X}^0 \in \Re^{n_1 \times m}$ required to build an initial PCA model for a few datasets, required to stabilize the algorithm. Here n_1 is the number of samples, expressed as rows and m is the number of variables, expressed as columns of the data matrix. The mean of each column is given by the vector: $\mu_1 = \frac{1}{n_1}(\mathbf{X}^0)^T I_{n_1}$, where $I_{n_1} = \begin{bmatrix} 1 & 1 & \dots & 1 \end{bmatrix}^T \in \Re^{n_1}$. Under special circumstances, for structural systems, data generally evolves from zero mean processes. However, as the nature of the data is not pre-assumed for deriving the recursive update expressions, it is centered to zero by subtracting its mean and scaled to a unit variance, according to: $\mathbf{X}_1 = \left(\mathbf{X}^0 - I_{n_1}\mu_1^T\right)\mathbf{\Sigma}_1^{-1}$ where, $\mathbf{\Sigma}_1 = diag(\sigma_{1.1},...,\sigma_{1.m})$, whose i^{th} element is the standard deviation of the ith measured variable ($i = 1,...,m$). The basic covariance matrix can be obtained as: $\mathbf{\Omega}_1 = \frac{1}{n_1-1}\mathbf{X}_1^T\mathbf{X}_1$. It can be expected that the new block of data will augment the data matrix and calculate the covariance matrix recursively, as and when a newer set of data streams in, that forms the basis of the FOP approach. Considering the kth sample, it is assumed that μ_k, $\mathbf{X_k}$ and $\mathbf{\Omega_k}$ are calculated. The next step of the algorithm is to implement the updates for the $(k+1)^{th}$ sample point, recursively.

Considering, $\mathbf{X}_{k+1}^0 = \left[\begin{array}{cc}\mathbf{X}_k^0 & \mathbf{X}_{n_{k+1}}^0\end{array}\right]^T$, for all the $k+1$ sample points, the mean vector μ_{k+1} is related to the mean vector at the kth sample point by the following expression:

$$\mu_{k+1} = \frac{N_k}{N_{k+1}}\mu_k + \frac{I_{n_{k+1}}}{N_{k+1}}\left(\mathbf{X}_{n_{k+1}}^0\right) \tag{4.1}$$

where, the quantity $N_k = \left(\sum_{i=1}^{k} n_i\right)$. The recursive calculation for the data matrix at the $(k+1)^{th}$ sample is given by:

$$\begin{aligned}\mathbf{X}_{k+1} &= \left(\mathbf{X}_{k+1}^0 - I_{k+1}\mu_{k+1}^T\right)\mathbf{\Sigma}_{k+1}^{-1} \\ &= \left(\begin{array}{c}\left(\mathbf{X}_k^0 - I_k\mu_k^T\right)\mathbf{\Sigma}_k^{-1}\mathbf{\Sigma}_k\mathbf{\Sigma}_{k+1}^{-1} - I_k\Delta\mu_{k+1}^T\mathbf{\Sigma}_{k+1}^{-1} \\ \left(\mathbf{X}_{n_{k+1}} - I_{n_{k+1}}\mu_{k+1}^T\right)\mathbf{\Sigma}_{k+1}^{-1}\end{array}\right) \\ &= \left(\begin{array}{c}\mathbf{X}_k^0\mathbf{\Sigma}_{k+1}^{-1} - I_k\mu_k^T\mathbf{\Sigma}_{k+1}^{-1} - I_k\Delta\mu_{k+1}^T\mathbf{\Sigma}_{k+1}^{-1} \\ \mathbf{X}_{n_{k+1}}\mathbf{\Sigma}_{k+1}^{-1} - I_{n_{k+1}}\mu_{k+1}^T\mathbf{\Sigma}_{k+1}^{-1}\end{array}\right)\end{aligned} \tag{4.2}$$

where, $\mathbf{\Sigma}_j = diag(\sigma_{j.1},...,\sigma_{j.m}), j = k, k+1$ and $\Delta\mu_{k+1} = \mu_{k+1} - \mu_k$. Following similar lines of development, it is easy to recognize that the recursive calculation of the covariance matrix has the following form [8]:

$$\begin{aligned}(N_{k+1}-1)\mathbf{\Omega}_{k+1} = \mathbf{X}_{k+1}^T\mathbf{X}_{k+1} - (N_k-1)\mathbf{\Sigma}_{k+1}^{-1}\mathbf{\Sigma}_k\mathbf{\Omega}_k\mathbf{\Sigma}_k\mathbf{\Sigma}_{k+1}^{-1} + \\ N_k\mathbf{\Sigma}_{k+1}^{-1}\Delta\mu_{k+1}\Delta\mu_{k+1}^T\mathbf{\Sigma}_{k+1}^{-1} + \mathbf{X}_{n_{k+1}}^T\mathbf{X}_{n_{k+1}}\end{aligned} \tag{4.3}$$

For data evolving from structural systems, the recursive update of the sample covariance matrix requires only rank-one modification. For this study, the covariance update is updated at each sample point, instead of updating the entire model, thereby relatively reducing the associated computational effort, for real time monitoring of structures. The recursive estimation shown in Eq. (4.3) can be reduced to:

$$\mathbf{\Omega}_{k+1} = \frac{k}{k+1}\mathbf{\Sigma}_{k+1}^{-1}\mathbf{\Sigma}_k\mathbf{\Omega}_k\mathbf{\Sigma}_k\mathbf{\Sigma}_{k+1}^{-1} + \mathbf{\Sigma}_{k+1}^{-1}\Delta\mu_{k+1}\Delta\mu_{k+1}^T\mathbf{\Sigma}_{k+1}^{-1} + \frac{1}{k+1}\mathbf{X}_{k+1}\mathbf{X}_{k+1}^T \tag{4.4}$$

For structural systems in general, data evolves from zero mean processes. Scaling the data to a unit variance, Eq. (4.4) can be rewritten as:

$$\mathbf{\Omega}_{k+1} = \frac{k}{k+1}\mathbf{\Omega}_k + \frac{1}{k+1}\mathbf{X}_{k+1}\mathbf{X}_{k+1}^T \tag{4.5}$$

Considering the EVD of the covariance matrix to be of the form $\mathbf{E}_{k+1}\mathbf{\Lambda}_{k+1}\mathbf{E}_{k+1}^T$, where $\mathbf{E}$ and $\mathbf{\Lambda}$ denote the orthonormal eigenvector and diagonal eigenvalue matrices respectively, with $\alpha_{k+1} = \mathbf{E}_k^T\mathbf{X}_{k-1}$, the following recursive formula is obtained, on substitution in Eq. (4.5):

$$\mathbf{E}_{k+1}(k+1)\mathbf{\Lambda}_{k+1}\mathbf{E}_{k+1}^T = \mathbf{E}_k[k\mathbf{\Lambda}_k + \alpha_{k+1}\alpha_{k+1}^T]\mathbf{E}_k^T \tag{4.6}$$

For a finitely large sample size, the term $[k\mathbf{\Lambda}_k + \alpha_{k+1}\alpha_{k+1}^T]$ is strongly diagonally dominant, which allows the application of Gershgorin's theorem [8], establishing the fact that while its eigenvalues will retain a structure close to the diagonal portion $[k\mathbf{\Lambda}_k]$, the corresponding eigenvectors will be close to identity [8].

In order to intuitively understand FOP, consider the eigenvalues of a perturbed matrix $\mathbf{C}+\mathbf{\Delta C}$ to be of the form $\mathbf{\Lambda}+\alpha\alpha^T$, i.e., the *rank-one update* of the matrix $\mathbf{\Lambda}$. Using the following definitions:

$$\begin{aligned} \mathbf{CV} &= \mathbf{\Lambda V} \\ (\mathbf{C}+\mathbf{\Delta C})(\mathbf{V}+\Delta_{\mathbf{V}}) &= (\mathbf{V}+\Delta_{\mathbf{V}})(\mathbf{\Lambda}+\Delta_{\mathbf{\Lambda}}) \end{aligned} \tag{4.7}$$

where, $\Delta_{\mathbf{V}}$ and $\Delta_{\mathbf{\Lambda}}$ are the perturbation matrices. The EVD of the diagonally dominant term, can be expanded as follows:

$$\mathbf{C}+\mathbf{\Delta C} = \mathbf{V\Lambda V}^T + \mathbf{V\Lambda}\Delta_{\mathbf{V}}{}^T + \mathbf{V}\Delta_{\mathbf{\Lambda}}\mathbf{V}^T + \Delta_{\mathbf{V}}{}^T\mathbf{\Lambda V}^T + O(\epsilon^3) \tag{4.8}$$

Recognizing, that $\mathbf{C} = \mathbf{V\Lambda V}^T$ and invoking the fact that $\mathbf{VV}^T = \mathbf{I}$, the EVD of the perturbed matrix $\mathbf{\Delta C}$ by using Eq. 4.8 and ignoring second order perturbation terms can be written as:

$$\mathbf{\Delta C} = \alpha\alpha^T = \mathbf{V\Lambda}\Delta_{\mathbf{V}}{}^T + \mathbf{V}\Delta_{\mathbf{\Lambda}}\mathbf{V}^T + \Delta_{\mathbf{V}}{}^T\mathbf{\Lambda V}^T + O(\epsilon^3)$$

The above expressions provide a central idea to the data driven nature of the FOP approach. Based on the above concepts, the proposed FOP technique has been successfully applied for multivariate datasets. An analogy can be drawn where the structure is instrumented at multiple DOF and the full rank sensor data is utilized for recursive estimation of the covariance matrix. However, due to cost and other impeding factors, instrumentation of the structure at every DOF might not be possible; therefore, the recursive updates at each time stamp is carried out using a subset of input channels from the structure. For a special case, where there is only a single data channel available as input, a recursive update can be carried out by generating a Hankel matrix out of the time series and carrying out the above mentioned formulations. This method, popularly called RSSA, is discussed in the following chapters and hence not reported here in detail. The methodology of the recursive estimation for the different types of datasets are described next in detail.

4.3.1 RPCA: Theoretical development using POMs

As a combination of detailed theoretical derivation for the basic PCA methodology, let the LNMs be represented as $\mathbf{V}$. Let the POMs be represented by $\mathbf{W}$ which are expressed as a sum of LNMs and error terms. Therefore, the expression $\mathbf{W} = \mathbf{V} + \varepsilon$ holds good. Since the relation $\mathbf{X} = \mathbf{VQ}$ is valid, it can be inferred that $\mathbf{Q} = \mathbf{V}^\mathbf{T}\mathbf{X}$.

Therefore, the POMs can be expressed as:

$$\boldsymbol{\Psi} = \mathbf{W}^\mathbf{T}\mathbf{X} = \mathbf{Q} + \boldsymbol{\Gamma} \tag{4.9}$$

where $\mathbf{Q}$ is the modal ensemble matrix and Γ is the matrix containing the error terms. Substituting $\mathbf{X} = \mathbf{W}\boldsymbol{\Psi}$, the *covariance matrix of the physical responses* can be obtained as:

$$\mathbf{R} = \frac{1}{N}\mathbf{X}\mathbf{X}^\mathbf{T} = \frac{1}{N}\mathbf{W}[\mathbf{Q}+\boldsymbol{\Gamma}]\,[\mathbf{Q}+\boldsymbol{\Gamma}]^T\mathbf{W}^\mathbf{T} \tag{4.10}$$

Equation 4.10 expresses the POCs of the covariance matrix of the acceleration response $\mathbf{X}$ in terms of the LNMs and errors. The basic RPCA equation 4.11 can be written as:

$$\mathbf{R}_k = \frac{k-1}{k}\mathbf{R}_{k-1} + \frac{1}{k}\mathbf{X}_k\mathbf{X}_k^T \tag{4.11}$$

where $\mathbf{R}_k$ and $\mathbf{X}_k$ are the covariance matrix and the matrix of the data points at the kth instant, respectively; and $\mathbf{R}_{k-1}$ denotes the covariance matrix at the $(k-1)^{th}$ instant. The covariance estimate $\mathbf{R}_k$ can be expressed as an eigen decomposition as shown:

$$\mathbf{R}_k = \mathbf{W}_k\boldsymbol{\Omega}_k\mathbf{W}_k^T \tag{4.12}$$

Thus for the $(k-1)^{th}$ data point the eigenvalue decomposition of $\mathbf{R}_{k-1}$ can be expressed as, $\mathbf{R}_{k-1} = \mathbf{W}_{k-1}\boldsymbol{\Omega}_{k-1}\mathbf{W}_{k-1}^T$, and the gain depth parameter β_k can be defined as $\beta_k = \mathbf{W}_{k-1}^T\mathbf{X}_k$

On substituting the value of the gain depth parameter and the covariance estimate in Eq. 4.11, the following expressions can be obtained

$$\mathbf{W}_k(k\boldsymbol{\Omega}_k)\mathbf{W}_k^T = \mathbf{W}_{k-1}\{(k-1)\boldsymbol{\Omega}_{k-1} + \beta_k\beta_k^T\}\mathbf{W}_{k-1}^T \tag{4.13}$$

For the RPCA algorithm to be stable and robust, it is important that the term $\{(k-1)\boldsymbol{\Omega}_{k-1} + \beta_k \beta_k^T\}$ is diagonally dominant, which can be demonstrated by expanding $\boldsymbol{\Omega}_{k-1}$ in terms of LNMs and error terms as follows:

$$\boldsymbol{\Omega}_{k-1} = [\mathbf{Q}_{k-1}{}^{\mathbf{T}}\mathbf{Q}_{k-1} + \mathbf{Q}_{k-1}\boldsymbol{\Gamma}_{k-1}^T + \mathbf{Q}_{k-1}{}^{\mathbf{T}}\boldsymbol{\Gamma}_{k-1} + \boldsymbol{\Gamma}_{k-1}\boldsymbol{\Gamma}_{k-1}^T] \tag{4.14}$$

From equation 4.14, it is clear that $\boldsymbol{\Omega}_{k-1}$ can be understood as a sum of $\mathbf{Q}^{\mathbf{T}}\mathbf{Q}$ and the first order error terms [8]. As $N \to \infty$, $\mathbf{Q}^{\mathbf{T}}\mathbf{Q}$ is approximately diagonal for systems having mild to moderate damping under sufficiently broad band excitations. Equation 4.14 forms the basis of establishing diagonal dominance of the $\boldsymbol{\Omega}_{k-1}$ matrix. Gershgorin's theorem can now be applied on the diagonally dominant matrix which provides recursive eigen space updates using perturbation techniques at each point in time. For dynamical systems of different order (such as chemical systems), [19] the above equations would not hold true and the concept of diagonal dominance has to be enforced, for the application of Gershgorin's theorem. Hence for a structural system, the recursive eigen space update is obtained using a first order perturbation (FOEP) approach [8, 10], which provides a less computationally intensive algorithm in a recursive framework for the eigen value decomposition $(\mathbf{T}_k \boldsymbol{\Lambda}_k \mathbf{T}_k^T)$ of the term $(k-1)\boldsymbol{\Omega}_{k-1} + \beta_k \beta_k^T$, yielding the following iterative update equations:

$$\left.\begin{aligned} \mathbf{W}_k &= \mathbf{W}_{k-1}\mathrm{T}_k \\ \boldsymbol{\Omega}_k &= \frac{\boldsymbol{\Lambda}_k}{k} \end{aligned}\right\} \tag{4.15}$$

Equation 4.15 provides an iterative relation between eigen spaces at consecutive time instants. On using the FOEP approach, the recursive eigen vectors obtained at each time instant are not ordered in the same sequence as the previous time instant, thus presenting the problem of permutation ambiguity [23]. This can be resolved by arranging the basis vectors according to the decreasing order of the corresponding eigenvalues in $\boldsymbol{\Omega}_{\mathbf{k}}$.

4.4 Real time condition indicators

A condition indicator (CI) is a statistical feature or quantity extracted from a monitored system data that provides information about the condition of the structure. Detecting damage patterns is one of the primary topics addressed in the SHM literature [5, 6, 10, 16] – more so in recent times – where real-time damage detection has gained prominence over the last few years [8–13]. A proper selection of CIs allow pattern recognition algorithms to indicate damage through changes in its patterns that are clear from visual inspections and graphical representations. On the other hand, if features that are employed do not provide any correlation to the damage event, the most clever pattern recognition algorithm will not significantly improve the damage detection process. In this context, a summary of the terminologies associated with employing CIs are provided in Table 4.3 below.[1]

Table 4.3: Feature classification terminologies.

Feature extraction	**Feature selection**
Transformation of extracted data into an alternative form or space where correlation with damage is more easily observed	Selection of appropriate feature for detecting damage
Usually model-based, physics-based or data-based approach	Features selected should be unaltered under operational and environmental variability
Alternatively, impedance-based and wave-propagation based features can directly compare with data waveforms	Features sensitive to damage are usually sensitive to changes in system response

A common method of feature selection is based on applying engineered flaws – similar to the ones expected in operating conditions – to develop an inital estimate of the parameters that lead to damage. Regardless of the nature of the system, the data extracted from the monitored system can be effectively used to distinguish between the healthy and damaged states by a thorough scrutiny corresponding to each time point. This forms the basics of online damage detection facilitated by RPCA. The present study focusses on recursive analysis of streaming data that yields online updates of eigenvalues and eigenvectors. An important consideration to note here is that the eigenspace by itself is not reflective of damage patterns unless processed by a set of condition indicators. Several damage detection and SHM techniques have been proposed in the literature [3, 5, 10, 16] that involve use of specific CIs whose changes corroborate to the damage in the system. The difficulty in choosing and utilizing CIs for the present framework arises from the fact that most of their traditional counterparts are not amenable to online implementation. The process therefore employs a set of recursive CIs that are responsive towards online implementation. These indicators are based on the deviation in eigenspace due to damage which is manifested through changes in the visualization of the CIs. Following this train of thought, temporal recursive residual errors (χ_{RR}), spatial recursive residual errors (ε_{RR}), eigen-vector change (EVC) and scatter plots have been presented as the chosen CIs for this study.

[1]One must not confuse a *feature* with a *metric*. Technically, a metric is a selected quantity that defines the similarity (or dissimilarity) between two features of interest. The general structure of a metric can be visualized as a distance function – such as the Euclidean or the Mahalanobis distance – that forms the basis of correlation between two datasets. On the other hand, a feature can be extracted from a single dataset, much opposed to a metric, which quantifies the patterns between two or more datasets.

4.4.1 Recursive residual error (RRE)

One of the primary CIs used for the present study is the recursive residual error (RRE). As will become more evident in the subsequent sections, RRE has proven to be the most robust and least outlier prone of all the discussed CIs. RRE comes in two flavors: temporal or global RRE (χ_{RR}) and spatial or local RRE (ε_{RR}). The main motivation for this CI is derived from the use of residual error as a criterion in quantification of nonlinear behavior [24] which presents its use as a measure of distortion of subspaces for a nonlinear system with an increase in levels of excitation. The RRE as utilized in the present work is derived as follows.

4.4.1.1 Temporal RRE

Let $\mathbf{W}_k$ be a matrix of eigenvectors computed at the kth time instant and $\mathbf{\Lambda_k}$ be the corresponding diagonal matrix of eigenvalues at that instant. The eigenvalues are structured in descending order of magnitude, and the corresponding eigenvectors are re-arranged. Let, $\mathbf{W_k}$= $[\mathbf{W_k^1 W_k^2}]$ such that, $\mathbf{W_k^1}$ is said to represent the least number of eigenvectors whose corresponding eigen values explain more than 90% of the variance. Considering a damage at the end of the $(k-1)^{th}$ instant, the subspace spanned by the updated eigenvector $\mathbf{W_k^1}$ deviates in comparison to the subspace spanned by eigen vectors at the previous time stamp $\mathbf{W_{k-1}^1}$. Apart from the instances of damage, or, the initial few seconds, it can be safely assumed that $\mathbf{W_k^1} \cong \mathbf{W_{k-1}^1}$, as there is no significant deviation in the eigenspace otherwise. Based on this assumption, the RREs due to projection of the response at a particular time instant k onto $\mathbf{W_{k-1}^1}$ is evaluated as:

$$\mathbf{X}^*(k) = \mathbf{W}_{k-1}^1 * \mathbf{W}_k^{1^T} * \mathbf{X}(k) \tag{4.16}$$

Based on the above concepts, for detecting the instant of damage, the RREs proposed here are χ_{RR-1} and χ_{RR-2} which are expressed by the following equations:

$$\left.\begin{aligned} \chi_{RR-1} &= \left\|\mathbf{X}^*(k) - \mathbf{W}_k^{1^T} * \mathbf{X}(k)\right\|^2 \\ \chi_{RR-2} &= \|\mathbf{X}^*(k) - \mathbf{X}(k)\|^2 \end{aligned}\right\} \tag{4.17}$$

From the Eq. 4.17, χ_{RR-1} can be interpreted as the distance metric between the transformed response and its projection on the subspace at the previous time stamp. Similarly, χ_{RR-2} provides a measure of the difference between projections of the transformed response and the original response data.

4.4.1.2 Local RRE

Identifying the damage instant simultaneously with locating the damage in a single framework is more involved – especially if attempted in a recursive construct

– towards online damage detection. Using CIs that are amenable to identifying changes in the local neighborhood of damage, the RPCA-RRE algorithm provides simultaneous detection-localization information for a monitored system. Generally speaking, damage localization in an online framework requires modifications in RRE formulation. **First**, local RREs need to be expressed corresponding to each DOF (ε_{RR}) – unlike the temporal (global) RREs (χ_{RR}) that were calibrated to the entire system. The key assumption is that when a column of a MDOF structure is damaged, its effect will be pronounced in the neighboring DOFs, a process that in turn is manifested as a change in the local RREs for that particular DOF. While finding out the instant of damage (Eq. 4.17), the concept of spatiality is lost as only global RREs are defined. This motivates the need for a new set of RREs (local or spatial RREs (ε_{RR})) that preserve the individual contribution of the responses, utilized to find the exact location of damage. From Eq. 4.16 it is clear that the projections of the transformed responses $\mathbf{X}^*(k)$ and the actual responses $\mathbf{X}(k)$ can be expressed as vectors with each individual element corresponding to a degree of freedom (m : total number of DOFs) according to: $\mathbf{X}^*(k) = [x_1^*(k), x_2^*(k), \ldots\ldots, x_m^*(k)]^T$ and $\mathbf{X}(k) = [x_1(k), x_2(k), \ldots\ldots, x_m(k)]^T$ where k represents the time instant at which the RRE is estimated. Consider each element of $\mathbf{X}^*(k)$ and $\mathbf{X}(k)$ be represented as $\mathbf{x_i}^*(k)$ and $\mathbf{x_i}(k)$ respectively, where (**i**) corresponds to a particular DOF. This yields a time series of RRE (labeled as $\varepsilon_{RR} - Y_i$) corresponding to the response for each DOF, which can be expressed as

$$\varepsilon_{RR} - Y_i = \left|\mathbf{x}_i^{*^2}(k) - \mathbf{x}_i^2(k)\right| \tag{4.18}$$

Second, spatial RREs are formulated to have an ensemble structure associated with them. This being, the average RRE corresponding to the ith response for K data points, is finally estimated according to,

$$\langle \varepsilon_{RR} - Y_i \rangle = \frac{\sum_{k=1}^{K} \left|\mathbf{x}_i^{*^2}(k) - \mathbf{x}_i^2(k)\right|}{K} \tag{4.19}$$

The above expression yields the local RRE averaged over K samples. This is particularly useful, as shown later, in quantifying the percentage change in RRE (an indirect measure of loss of stiffness) pre and post damage.

4.4.2 Recursive eigen vector change

From the precursory discussions, it is now evident that a significant distortion in the eigenspace characteristics at the onset of damage can be expected. Hence, tracking deviations in the eigenvector updates are expected to show changes that could potentially correspond to the instant of damage. Pivoted around this key concept, a new CI, eigenvector change (EVC) has been developed in the context of real time damage detection. The eigenspace updates at each instant of time are tracked and an accurate damage instant is identified when a significant distortion in the eigenspace is encountered. In the due course of rigorous inspection over a varied range of sample numerical simulations and experimental verifications, it was later found that this CI did not provide successful representation of the damage instant for quite a number of instances. As it turns out, unlike the RREs, EVC is highly susceptible to false detections. Erroneous detection often arises due to *ambient noise* or some other disturbances from the excitation data, that sometimes become inevitable during experimentation. The EVC as a CI was thus kept in consideration for a certain number of cases where it actually showed some definite results for the damage instant.

$$EVC = \sqrt{diag([\mathbf{W^1_k} - \mathbf{W^1_{k-1}}] * [\mathbf{W^1_k} - \mathbf{W^1_{k-1}}]^T)} \tag{4.20}$$

4.4.3 Outlier detection using correlation coefficient (ρ)

Outlier analysis is a thriving area of health monitoring methods banked mostly upon probability and statistics. A brief description of the basic analysis concepts is provided here – interested readers are referred to the seminal work by Sohn et al. [25] for more details.

Outliers can be defined as observations (or a subset of observations) which appear to be inconsistent with the remainder set of data. They can arise due to mechanical faults, changes in system behavior, human practice or instrument error. The discordancy of a candidate outlier can be compared against an objective function to be adjudged statistically likely or unlikely to have evolved from the monitored process. The use of scatter plots in the present context remains a tool for validating the findings from RRE. By no means, should a scatter plot be confused as a part of an online process – intrinsically, these plots are generated by accumulating data over a period of time, and are therefore, a *batch-mode* implementation. Scatter plots between two preselected transformed responses and their corresponding correlation coefficients (ρ) are calculated in the neighborhood of damage (small windows before and after the instant of damage). It is observed that there is considerable directional change in the scatter plots (corresponding to a change in the sign of ρ) only at the instant of damage, which serves as a viable indicator for removing outliers in the algorithm. A point to note here is – there is no change in the sign of the correlation coefficient *even* due to the presence of outliers. Damage manifested in the form of loss of stiffness is conveniently visualized from the RRE plots and is validated by the change in the sign of ρ.

4.5 Proposed algorithm

The overall methodology followed in the context of the RPCA algorithm is shown in Fig. 4.1, that comprises of two key ingredients: *temporal* and *spatial* damage detection modules that occur simultaneously. The first one entails processing of the acceleration data by the RPCA algorithm as it streams in real-time. As previously explained, the RRE is tracked online for any major and minor changes. Whenever a significant alteration is observed, the outlier detection step operates on the RRE data accumulated till the event instant (i.e., the instant of change) to determine whether the change corresponds to an outlier or a damage. Once the damage instant is detected, the spatial damage detection module further resolves the location of damage.

In a nutshell, the summary of the key steps involved in the RPCA-RRE method are enumerated below:

1. First, the initial estimate of the eigen vector and eigen value matrices are computed for a few initial data points.

2. The RPCA algorithm operates on the real time input of the streaming acceleration data.

3. Using the recursive gain depth parameter, the covariance estimate at the present time instant is derived using the covariance estimate at the preceding time instant as per Eqs. 4.11 to 4.13. From the recursive updates, the eigen vector and eigen value matrices (eigenspace) are updated using the FOEP approach (Eq. 4.15).

4. The recursive CIs (RRE) employed on the updated eigenspace from the RPCA algorithm are then monitored to observe any sudden changes indicating a possible damage (by setting a tolerance limit of 30% in the change of the RRE before and after damage). Changes in RRE greater than the tolerance limit are then verified using scatter plots. Once the instant of damage is corroborated using scatter plots, the algorithm proceeds to detect the *location of damage* using local RREs in the spatial mode.

5. For the cases of no detections, the present eigenspace is reverted back to the RPCA algorithm which takes it as input along with the new data set obtained at the next time instant, to process the new eigenspace.

A flowchart as shown in Fig. 4.1 outlines the proposed damage detection scheme. From Fig. 4.1, it is clear that RPCA algorithm can process data as and when it streams (online) and provides a recursive update of the eigen structure at each data point which is further utilized by the recursive CIs to track the instant of damage. Hence, it can be well established that the algorithm does not depend on a baseline

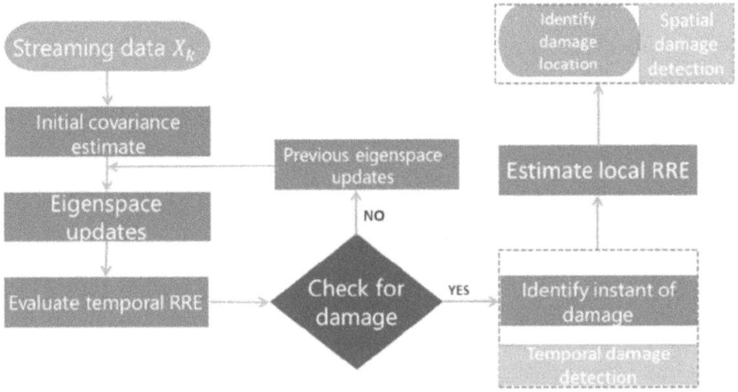

Figure 4.1: Flowchart for the RPCA algorithm.

(reference value) – unlike batch PCA, which is solely dependent on a batch of response data in order to detect damage. Cluster plots and RREs are utilized in tandem to detect the time of damage and for removal of outliers. An essential point to note is that there are no parameters controlling the functioning of the algorithm, i.e., it is fundamentally *parameter free*. However, it should be remembered that cluster plot representation is just a visual aid for validating the instant of damage obtained from RRE. The window size employed here is generally arbitrary but can be catered to specific problems depending upon the nature of the system and the damage to be detected. If there is no change in the orientation of the scatter plots (as a result of no change in the sign of the correlation coefficient) – where RRE shows a deviation – it can be safely assumed to be an outlier. Since the cluster plots are visualization tools, they consume computational resources, hence, should be used judiciously.

4.6 Numerical example

In order to illustrate an application of the proposed method, numerical simulations are performed on a 5 storey structure modeled with 4 floors and a Buoc-Wen (B-W) base oscillator. The simulated nonlinear change of state is *contextually defined* as damage. The structure is subjected to white noise excitation of a time duration of $50s$. The level of nonlinearity is controlled by a parameter κ, that is basically used as a multiplicative scale factor with the nonlinear force term. For the numerical simulations, two types of input excitations are used: stationary zero mean Gaussian white noise and an earthquake excitation (El Centro ground motion). For both cases, the input data are sampled at 50 Hz.

4.6.1 Structural model and simulation parameters

The 5 storey structure is modeled with 4 floors and a base. The LRB isolator separates the base from the surrounding ground. The state equations for this system subjected to an external excitation vector **w** can be written as:

$$
\begin{aligned}
x &= \mathbf{A}x + \mathbf{E}w \\
y &= \mathbf{B}x
\end{aligned}
\tag{4.21}
$$

Here, the vector **x** is the vector of states and the vector **y** represents the output vector, which is governed by the **B** matrix. The system matrix, **A**, and the excitation matrix **E** are given by

$$
\begin{aligned}
\mathbf{A} &= \begin{bmatrix} [\mathbf{O}]_{5\times5} & [\mathbf{I}]_{5\times5} \\ -\mathbf{M}^{-1}\mathbf{K} & -\mathbf{M}^{-1}\mathbf{C} \end{bmatrix} \\
\mathbf{E} &= \begin{bmatrix} 0 & 0 & 0 & 0 & 0 & -\frac{1}{m} & -\frac{1}{m} & -\frac{1}{m} & -\frac{1}{m} & -\frac{1}{m} \end{bmatrix}^T
\end{aligned}
\tag{4.22}
$$

The equation of motion for the system can be summarized as:

$$
\mathbf{M}\ddot{u} + \mathbf{C}\dot{u} + \mathbf{K}u = \mathbf{\Lambda} f - \mathbf{M}\mathbf{I}\mathbf{u_g}
\tag{4.23}
$$

Here, **M**, **C**, and **K** are the assembled mass, damping, and stiffness matrices, respectively. A simple shear building representation is assumed to arrive at the expressions for **M**, **C**, and **K** which are skipped here for brevity. Numerical simulations are carried out on a simple 5 DOF mass, spring, and dashpot system. The mass at each of the four floor levels from the top is 7461 kg and at the base is 6800 kg. The damping coefficients for each floor level above the base is 23.71 kNs/m and 3.74 kNs/m for the base. The stiffness coefficients for each of the floors above the base it is 11912 kN/m and that for the base is 232 kN/m.

In Eq. 4.23, $\mathbf{\Lambda}$ represents the location of the base at the point of application of the force due to the LRB base isolator and $\mathbf{u_g}$ represents the ground acceleration. The vector **x** represents the displacement of each floor including the base and f represents the nonlinear force exerted by the LRB. It should be noted that the forces due to base damping and stiffness terms (k_b and c_b) have been included in the force f due to the LRB base isolator, which can be expressed as:

$$
f = \kappa z Q_{pb} + k_b x_b + c_b \dot{x}_b
\tag{4.24}
$$

where, $Q_{pb} = \left(1 - \frac{k_{yield}}{k_{initial}}\right) Q_y$ and k_b and c_b are the stiffness and the viscous damping respectively, in the horizontal direction. $k_{initial}$ is the initial shear stiffness and k_{yield} is the post yield shear stiffness of the LRB. The nonlinear force exerted by the LRB is expressed using the B-W system. The evolutionary variable z is used to provide the hysteretic component of the horizontal force,

$Q_{hyst} = zQ_{pb}$. The variable z can be obtained by solving the following nonlinear differential equation:

$$\dot{z} = -\gamma z |\dot{x}_b| \left|z^{n-1}\right| - \beta \dot{x}_b |z^n| + A\dot{x}_b \tag{4.25}$$

where γ, β, A and n are the shape parameters of the hysteresis loop. For the current model, $A = \left(\frac{k_{yield}}{k_{initial}}\right)$, $\gamma = \beta$ and $n = 1$. The yield force Q_y is selected as 5% of the total weight of the building which gives $Q_y = 17800$ kg and pre-yield to post-yield stiffness ratio $\left(\frac{k_{yield}}{k_{nitial}}\right) = \frac{1}{6}$. For the present study of the model, $\gamma = \beta = 39.1$. The constant κ controls the nonlinearity introduced into the equation of motion of the system (through the nonlinear force term). For instance, a change in κ from 1 to 0.3 is conveniently assumed as a 70% change in nonlinear characteristics of the system. For all subsequent discussions, this model will henceforth be referred to as *5-DOF B-W system*.

4.6.2 Results for white noise input excitation

Temporal damage detection cases are studied first by sequentially changing the κ corresponding to preselcted percentage changes of the nonlinear parameter for the B-W model. A small outlier detection study is first performed which presents the shortcomings of EVC as an effective damage detector – primarily due to the problems of false detection. Spatio-temporal damage detection results are then described in detail, which sheds a light on the efficacy and the performance of the RPCA algorithm over its traditional counterpart (PCA). Finally, in an attempt to explore the applicability of the method towards more realistic case studies, detection results for an underdetermined system is provided. The potential of the RPCA method is clearly portrayed from the damage detection results considering the case where the number of instrumented sensors is lower than the number of physical degrees of freedom.

4.6.2.1 Temporal damage detection results

The vibration response of a few preselected storeys of the B-W system is shown in Fig. 4.2. The case study progresses with gradual shifts in nonlinearity by 25%, 30%, and 75% – all corresponding to a fixed time instant 31s. As raw acceleration plots do not convey meaningful information regarding the instant of damage based on the raw vibration signature. These signals are then processed by the RPCA algorithm that provides a transformed response at each time stamp. Damage is detected using two iterative condition indicators – EVC and RREs (χ_{RR-1} and χ_{RR-2}). Figure 4.3(a) and (b) signify the damage instant by a sudden change in the mean level at 31s from the start of the excitation. A quick verification of the same is evident from Fig. 4.3(c) where the spikes at 31s corroborate the damage instant.

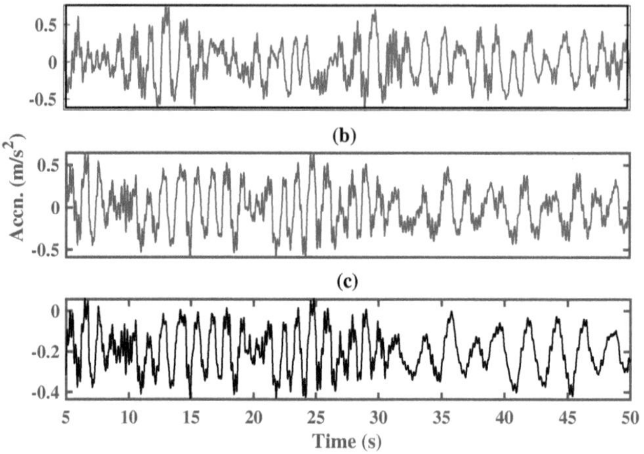

Figure 4.2: Output acceleration plots for $1st$, $3rd$ and $5th$ floor of the B-W system.

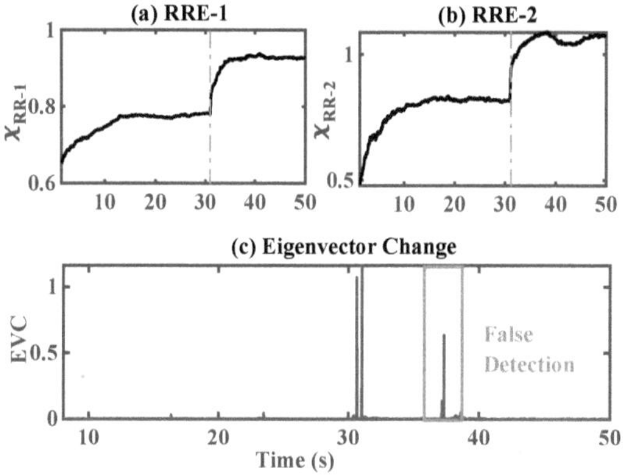

Figure 4.3: Sample detection studies using RREs and EVC.

However, due to the nature of the excitation (constituting a random process), the EVC method is prone to false detections as illustrated in the figure. The following detection studies are therefore, carried out by omitting EVC from the discussion.

By looking at Fig. 4.3(b) and 4.3(c), it can be seen that there is a significant shift in the estimate of the residual errors χ_{RR-1} and χ_{RR-2}. It can also be observed from Fig. 4.3(a), that EVC could capture the essence of fault detection which shows up as a significant peak at the damage instant. The EVC however, shows a period of activity around the 0-8s mark and also shows a false detection at 37s. This indicates that although EVC can be potentially applied for case-specific problems, its utility in most online damage detection studies can be compromised

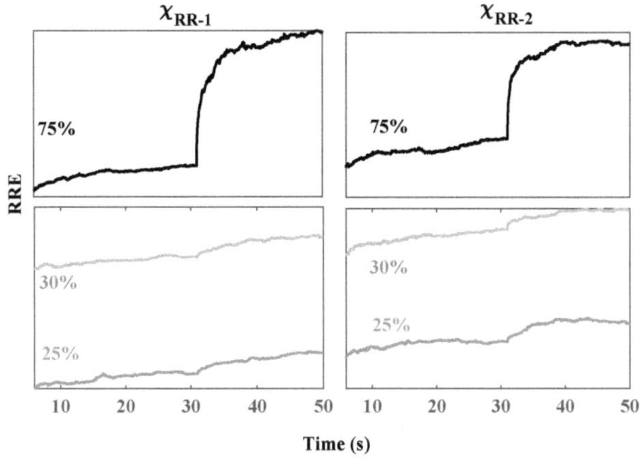

Figure 4.4: Damage detection using RREs for varying cases of non linearity change.

by false detections. Moving on, Fig. 4.4 shows the performance of χ_{RR-1} and χ_{RR-2} for various cases of damage. It is worth noting that the efficacy of RREs for less than 25% damage detection is slightly questionable which indicates that the current online framework is less reliable when the extent of damage suffered is low (i.e., less than 25%). It is therefore, suggested that the RPCA-RRE algorithms be employed in cases with long-term monitoring ensuring a higher percentage of structural damage.

The scatter plot between the transformed responses ($Y_t r^i$ vs $Y_t r^j$) can also serve as a robust visual CI for detecting damage. Since the transformed responses update recursively, showing point-wise scatter plots is time and memory consuming. To demonstrate the efficacy of scatter plots, data are considered in growing windows with an initial window size of 10s and at increments of 10s before damage. Therefore, each of the subplots shown in Fig. 4.5 represent windows of 10s from the start of excitation (i.e., 1st window comprises 0-10s, 2nd window corresponds to 10-20s, and so on). It can be clearly observed from the figure that there is a significant change in orientation of the scatter plot – which is also reflected by the change in the signs of the correlation coefficients – between the successive windows before and after damage. As damage instant at 31s is obtained from the RRE plot, the 4th window clearly corroborates the event for the specified level of non linearity.

Figure 4.6 demonstrates the robustness of RRE-1 for various damage scenarios – 30%, 50% and 75%. As observed from Fig. 4.6(c), the RRE shows a distinct shift in the mean level corresponding to a higher percentage change in non linearity. For lower percentages of damage, the changes are slightly less distinct and in some cases, negligible Fig. 4.6(a) and (b). However, this is not an impediment as far as the present online damage detection framework is considered, since

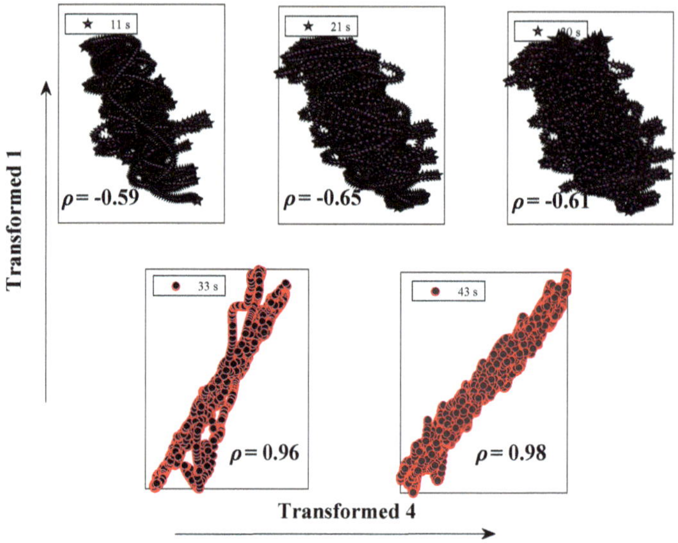

Figure 4.5: Scatter plot representation of damage.

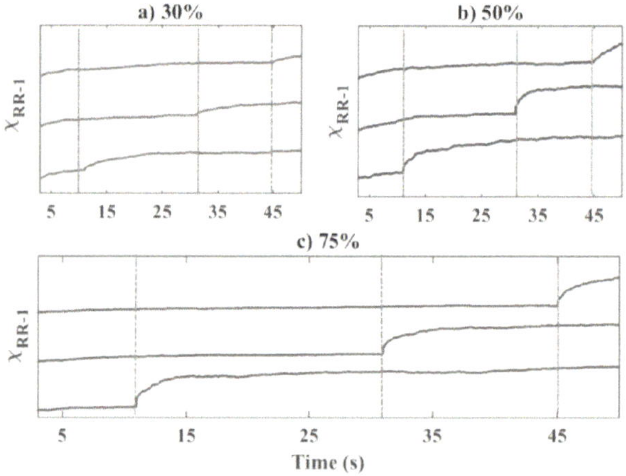

Figure 4.6: Detection results using RRE-1 for different damage cases.

any event of change can always be verified using scatter plots to finally pinpoint the occurrence of damage. The relative change in global RREs corresponding to different levels of damage is shown in Table 4.4. It is clear from the results in Table 4.4 that the percentage change in RRE increases with the level of damage.

Table 4.4: Global RREs for numerical modeling (using white noise).

Change in nonlinearity (%)	Pre-damage RRE	Post-damage RRE	% change
25.00	0.74	0.98	32.43
30.00	0.71	0.97	36.62
35.00	0.77	1.06	37.17
40.00	0.68	1.02	50.15
50.00	0.69	1.27	84.06

4.6.2.2 Spatial damage detection results

Numerical simulations of local damage are performed by varying the linear storey stiffness of an individual floor on the B-W model. The reader by now is familiar with the entire detection process – first, the acceleration response is processed by the RPCA algorithm and the eigenspace updates are tracked over time. Next, the eiegnspace is processed by the select set of CIs which finally reveal the damage instant. However, a crucial concept to recollect here is that the damage is *local* instead of *global* – and therefore, *local* RRE is employed to identify the location of structural damage in the system.

For the simultaneous temporal and spatial damage case (online spatio-temporal damage), 50% and 35% changes in linear stiffness of 3rd storey are considered. The nonlinearity level κ is set at 1 (fully nonlinear). Following a similar line of thought, one can arrive at Fig. 4.7 where the instant of damage is evident at 31s from the temporal RRE plot. Once the damage instant is detected, the spatial RRE $\varepsilon_{RR} - Y_i$ in a small neighborhood in the vicinity of damage (29s to 33s) are then investigated. The RRE plot corresponding to the third floor ($\varepsilon_{RR} - Y_3$) indicates a subtle change in the mean level which verifies the location of the simulated damage. In the same plot, however, the local RREs adopted for the remaining DOFs do not show any deviations from the mean level and are therefore, considered as undamaged – which corroborates the simulation study. In an attempt to investigate the performance of the algorithm for lower detection cases, spatial RRE for 35% damage is evaluated. From Fig. 4.8, it is evident that the spatial RRE for 35% damage indicates a visible change at t = $31s$. However, as the extent of stiffness degradation reduces, detecting spatial damage detection becomes increasingly difficult compared to the detection of the temporal damage alone.

4.6.3 Results for underdetermined case—white noise excitation

In this section, the possibility of applying the proposed algorithm to handle an under determined case of structural damage detection is explored. An under determined case can be defined as the instance in which the number of sensors

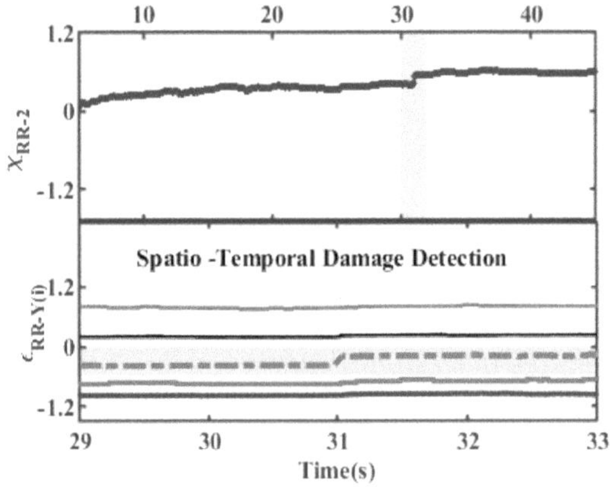

Figure 4.7: Spatio-temporal damage detection – 50% damage case.

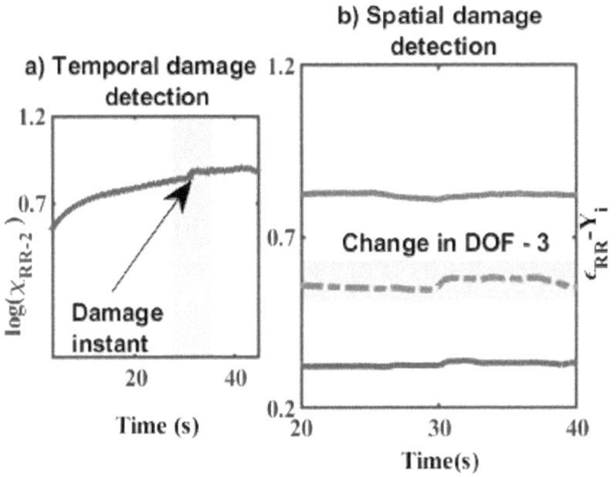

Figure 4.8: Spatio-temporal damage detection – 35% damage case.

instrumented is less than the number of physical DOF. Such a system might arise in flexible systems that are instrumented with a relatively smaller number of sensors because of cost considerations, unavailability of good quality sensors, improper accessibility – to name a few. So far the discussion has been limited to cases where full sensor measurements are available – which is a restrictive assumption – especially while dealing with practical problems in engineering. The proposed algorithm tries to address this issue by assuming that only a subset of the sensor measurements are available for online damage detection. In this context, the recursive CIs are explored to cast the problem of underdetermined mixtures within the framework of the RPCA method, as explained subsequently.

In order to conduct the damage identification study, it is assumed that the number of DOF in the structure equals the number of available measurements.

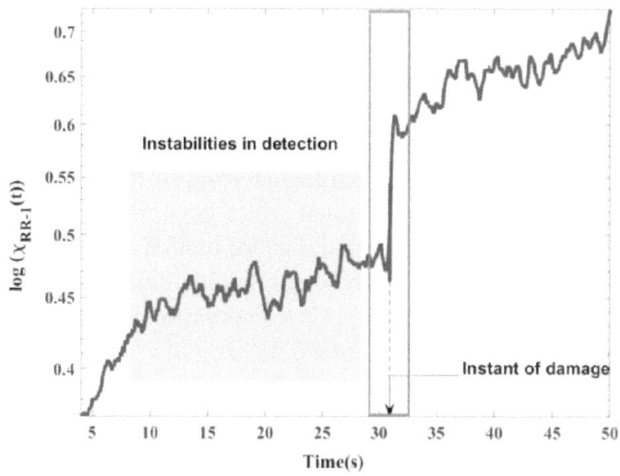

Figure 4.9: Detection results for the underdetermined case study.

Theoretically, to satisfy the rank condition of the eigenvector matrix, the number of instrumented DOF should be at least equal to number of actively participating modes. Hence for a 5-DOF structure, if only 3 degrees of freedom are instrumented with sensors, the proposed algorithm assumes the system to be a 3-DOF structure, yielding the corresponding eigenspace. The RPCA algorithm is applied on the streaming data which produces an eigenspace corresponding to the reduced DOF. Recursive CIs applied on the reduced eigenspace provides *visual aids* of damage manifestation in the system, as shown in Fig. 4.9.

To illustrate the applicability of the algorithm towards an underdetermined system, the model proposed in Section 4.6.1 with acceleration response data missing from the 2nd floor and the 4th floor are taken for analysis. Global damage was induced by a 25% change in nonlinearity at 31s. Real-time data streaming from all the sensors was made available to the RPCA algorithm as input for online damage detection and the results providing a case study of the RRE plot is shown in Fig. 4.9.

Figure 4.9 shows that the proposed algorithm is able to ascertain the instant of damage even without access to the complete set of output responses. As observed from the RRE plot, it is worth noting the fact that there is a period of activity before the actual instant of damage is reached – solely attributed to the instabilities arising from the underdetermined nature of the response. Despite these instabilities, the RPCA-RRE approach indicates the exact instant of damage through a change in the mean level of the plot. This substantiates the fact that the algorithm is fully equipped to tackle a practical situation that closely emulates a real-life scenario where the number of sensors instrumented in the structure are lesser than the physical DOFs.

4.6.4 Comparative study with batch PCA

With the discussions centred mostly towards RPCA, it becomes necessary to evaluate its performance against its batch counterpart – the traditional PCA algorithm. Considering the same B-W model used in the previous case studies, damage is induced by changing the nonlinearity level of the system by 50% at 31s. The model is excited using a Gaussian white noise input of 50s duration. While χ_{RR-1} still remains a potential candidate for recursive CI, the batch mode PCA would therefore require a similar indicator that identifies the damage instant in the windows. Therefore, an *analogous CI* for operation in batch mode $\chi^{b_i}_{RR-1}$ is defined to justify the comparison with RPCA.

At each time stamp, RPCA evaluates the χ_{RR-1} values that are tracked in real-time. On the contrary, batch PCA accumulates the data gathered over a definite period of time (in 10s windows) and calculates $\chi^{b_i}_{RR-1}$ using windowed data. The implementation of batch PCA over the data provides a single value of RRE for each 10s window span. The analogous batch RRE is calculated for any i^{th} window using traditional PCA through the following expression:

$$\chi^{b_i}_{RR-1} = \left\| \mathbf{X}^{b_i *} - \mathbf{W}^{b_i} \mathbf{X}^{b_i} \right\|^2 \tag{4.26}$$

In the above expression, $\mathbf{X}^{b_i}$ represents the response matrix containing the ith window of data. $\mathbf{W}^{b_i}$ is the orthogonal eigenvector matrix derived from the EVD of the covariance matrix $\frac{1}{N}\mathbf{X}^{b_i}\mathbf{X}^{b_i T}$ for the corresponding i^{th} window data matrix having N time instants. Additionally, the recursive CI (χ_{RR-1}) values are plotted against time to demonstrate the detection results based on the RPCA algorithm. In the same plot, the batch PCA detection results are plotted alongside for comparison.

From Fig. 4.10, it is observed that the RPCA-RRE method instantaneously detects the exact damage instant, through a significant change in the variation of χ_{RR-1} at $31s$. The batch PCA results using the CI $\chi^{b_i}_{RR-1}$ show almost constant values for the first three windows (prior to the instant of damage) – a fact also established through RPCA. A significant change in the value of $\chi^{b_i}_{RR-1}$ indicates that the damage instant lies somewhere in the 4th *window*, within the time interval between $30s$ to $40s$. However, the batch PCA based CI is unable to identify the exact instant of damage, as it indicates a *range* of time intervals where the probable damage instant might have occurred. Therefore, it becomes clear that windowing of data necessary for executing batch PCA-based damage detection prevents any possibility of its online implementation. The above results indicate that RPCA-RRE is advantageous over batch PCA in terms of exact identification of damage time instants and easy online applicability without the need for *ad-hoc windowing* of the response data.

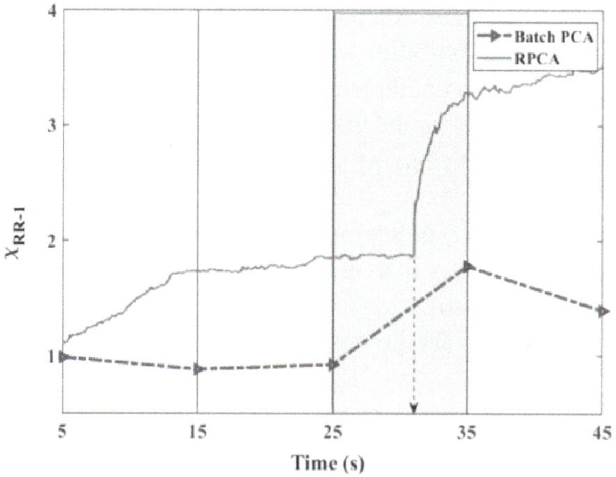

Figure 4.10: RPCA vs batch PCA – damage detection.

4.6.5 Results for El Centro ground excitation

After a satisfactory performance of the RPCA algorithm in detecting damage under ambient excitation conditions, it is natural to seek an investigation into its performance under a nonstationary regime. Towards this, the B-W model is excited using a scaled El Centro vibration data. The damage is simulated by changing the value of κ from 1 to 0.6 – corresponding to a 40% change in nonlinearity at 29s from the start of the excitation. The output vibration response is processed by the RPCA algorithm and the damage markers are employed to accurately identify the event. The detection results for this case are illustrated in Fig. 4.11. It is clear that χ_{RR-2} clearly shows significant change at 29s. From the

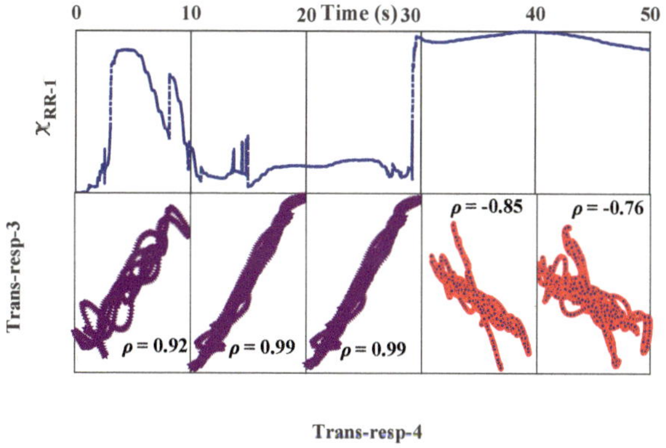

Figure 4.11: Detection results for El Centro input excitation.

RRE plot, it is evident that there is a period of activity before the actual damage instant. To eliminate the possibility of any false detections, scatter plots are employed and the correlation coefficient (ρ) corresponding to each 10s window is calculated. The scatter plots for the first three windows neither indicate a change in orientation nor mark a change in the sign of ρ. Progressing to the next window, the change in orientation is clearly visible from the figure. This is a result of the change in the sign of ρ, thereby indicating a damage event prior to this dataset. An important consideration here is that simultaneous spatial and temporal damage detection is difficult in this case owing to amplitude non-stationarity of the ground motion and therefore, the results are not reported.

References

[1] Behmanesh, I. and Moaveni, B. (2015). Probabilistic identification of simulated damage on the Dowling Hall footbridge through Bayesian finite element model updating. Structural Control and Health Monitoring, 22(3): 463–483.

[2] An, Y. and Ou, J. (2013). Experimental and numerical studies on model updating method of damage severity identification utilizing four cost functions. Structural Control and Health Monitoring, 20(1): 107–120.

[3] Wang, S. Q. and Li, H. J. (2012). Assessment of structural damage using natural frequency changes. Acta Mechanica Sinica, 28(1): 118–127.

[4] Gharibnezhad, F., Mujica, L. E. and Rodellar, J. (2015). Applying robust variant of principal component analysis as a damage detector in the presence of outliers. Mechanical Systems and Signal Processing, 50: 467–479.

[5] Li, Y. Y. and Chen, Y. (2013). A review on recent development of vibration-based structural robust damage detection. Structural Engineering and Mechanics, 45(2): 159–168.

[6] Deraemaeker, A., Reynders, E., De Roeck, G. and Kullaa, J. (2008). Vibration-based structural health monitoring using output-only measurements under changing environment. Mechanical Systems and Signal Processing, 22(1): 34–56.

[7] Pandey, A. K., Biswas, M. and Samman, M. M. (1991). Damage detection from changes in curvature mode shapes. Journal of Sound and Vibration, 145(2):321–332.

[8] Hearn, G. and Testa, R. B. (1991). Modal analysis for damage detection in structures. Journal of Structural Engineering, 117(10): 3042–3063.

[9] Kim, J. T., Ryu, Y. S., Cho, H. M. and Stubbs, N. (2003). Damage identification in beam-type structures: frequency-based method vs mode-shape-based method. Engineering Structures, 25(1): 57–67.

[10] Jaishi, B. and Ren, W. X. (2005). Structural finite element model updating using ambient vibration test results. Journal of Structural Engineering, 131(4): 617–628.

[11] Krishnan, M., Bhowmik, B., Tiwari, A. K. and Hazra, B. (2017). Online damage detection using recursive principal component analysis and recursive condition indicators. Smart Materials and Structures, 26(8): 085017.

[12] Bhowmik, B., Tripura, T., Hazra, B. and Pakrashi, V. (2019). First-order eigen-perturbation techniques for real-time damage detection of vibrating systems: theory and applications. Applied Mechanics Reviews, 71(6).

[13] Krishnan, M., Bhowmik, B., Hazra, B. and Pakrashi, V. (2018). Real time damage detection using recursive principal components and time varying auto-regressive modeling. Mechanical Systems and Signal Processing, 101: 549–574.

[14] Bhowmik, B., Krishnan, M., Hazra, B. and Pakrashi, V. (2019). Real-time unified single-and multi-channel structural damage detection using recursive singular spectrum analysis. Structural Health Monitoring, 18(2): 563–589.

[15] Bhowmik, B., Tripura, T., Hazra, B. and Pakrashi, V. (2020). Real time structural modal identification using recursive canonical correlation analysis and application towards online structural damage detection. Journal of Sound and Vibration, 468: 115101.

[16] Bhowmik, B., Tripura, T., Hazra, B. and Pakrashi, V. (2020). Robust linear and nonlinear structural damage detection using recursive canonical correlation analysis. Mechanical Systems and Signal Processing, 136: 106499.

[17] Farrar, C. R. and Worden, K. (2007). An introduction to structural health monitoring. Philosophical Transactions of the Royal Society A: Mathematical, Physical and Engineering Sciences, 365(1851): 303–315.

[18] Doebling, S. W., Farrar, C. R. and Prime, M. B. (1998). A summary review of vibration-based damage identification methods. Shock and Vibration Digest, 30(2): 91–105.

[19] Misra, M., Yue, H. H., Qin, S. J. and Ling, C. (2002). Multivariate process monitoring and fault diagnosis by multi-scale PCA. Computers & Chemical Engineering, 26(9): 1281–1293.

[20] Worden, K., Manson, G. and Allman, D. (2003). Experimental validation of a structural health monitoring methodology: Part I. Novelty detection on a laboratory structure. Journal of Sound and Vibration, 259(2): 323–343.

[21] Manson, G., Worden, K. and Allman, D. (2003). Experimental validation of a structural health monitoring methodology: Part II. Novelty detection on a Gnat aircraft. Journal of Sound and Vibration, 259(2): 345–363.

[22] Lämsä, V. and Raiko, T. (2010, August). Novelty detection by nonlinear factor analysis for structural health monitoring. In 2010 IEEE International Workshop on Machine Learning for Signal Processing (pp. 468–473). IEEE.

[23] Sadhu, A. and Hazra, B. (2013). A novel damage detection algorithm using time-series analysis-based blind source separation. Shock and Vibration, 20(3): 423–438.

[24] Hot, A., Kerschen, G., Foltête, E. and Cogan, S. (2012). Detection and quantification of non-linear structural behavior using principal component analysis. Mechanical Systems and Signal Processing, 26: 104–116.

[25] Sohn, H., Farrar, C. R., Hunter, N. F. and Worden, K. (2001). Structural health monitoring using statistical pattern recognition techniques. J. Dyn. Sys., Meas., Control, 123(4): 706–711.

Chapter 5

Multi-sensor Real Time Damage Detection Techniques (B): RPCA-TVAR

Generally, a typical monitoring setup strongly relies on acquisition from multiple measurement points – for free or forced vibrations – with or without the knowledge of the system excitation and properties. Vibration characteristics of civil infrastructure often change due to environmental effects (temperature, humidity and others) or operational conditions (changes in loading patterns, human activities, extraneous noise and more). From a physical viewpoint, such systems are commonly encountered in practice, and include earthquake-induced vibrations, vibrations of surface machineries, airborne structures, rotating machinery components of sea vessels, and so on. Mathematically, modeling and analysis of these systems is based on signal processing techniques considering time-varying and non-stationary effects of the system responses. This has received significant attention from the research community and is very important in a wide variety of applications, including SHM.

A very attractive and efficient field of analysis is to model the acquired structural response into time varying models. Towards this, the present chapter is dedicated to the theoretical development of RPCA using principal orthogonal components (POCs) of streaming response signals. Real-time detection analysis is primarily carried out using time-varying auto-regressive (TVAR) models that are well-equipped to identify the behavioural patterns of structural damage. This method extends the concepts of the RPCA based approach to determine the spatio-temporal damage patterns under nonstationary input excitations that give rise to varying structural patterns over time. The strategies discussed in this chapter are presented in an increasing order of sophistication. The fundamental concepts of the RPCA-TVAR approach, central to the development of the proposed methodology, are reviewed first. The fundamentals of TVAR modeling are presented next and are highlighted in the light of SHM applications in general, and damage detection in particular. As discussed in the previous chapter, the design of an appropriate monitoring system begins during the operational stage of the ideal SHM process. Since a detailed summary of all of these systems is beyond the scope of this book, the approach taken in this chapter will be to summarize the need for multiple sensing locations in a monitoring system and corresponding numerical simulations on the previously discussed Buoc-Wen (B-W) model will be highlighted. To examine the efficacy of the proposed methodology, experimental setups devised under controlled laboratory environments will be finally presented to verify structural damage instants in real-time. This will provide ground evidence to suggest the adaptability and robustness of the discussed method for practical applications. To facilitate easy understanding of the material in the subsequent discussions a list of acronyms is provided in Table 5.1.

Table 5.1: Important acronyms.

AR	Auto regressive
TVAR	Time Varying Auto Regressive
POC	Principal Orthogonal Components
DSF	Damage Sensitive Feature
HOM	Higher Order Moment

5.1 Motivation

As with SHM, damage detection has application in almost all engineered structures and mechanical systems including civil infrastructure, defence systems, power plants, commercial aerospace systems, and so on. Industrial revenue and marketing profits are realized when the ability to predict a damage and assess it in real-time requires minimal corrective action and mitigation. With such predictive operations in place, real-time fault detection enables owners to schedule maintenance in advance, enforce periodic downtimes, and optimize hardware for leasing – thereby maximizing the revenue-generating potential of these assets.

Online fault detection for such systems can be complicated – especially if the nature of the system excitation is unknown – which usually leads to auxiliary computational expenses that require the skills of experienced practitioners for analysis. To alleviate the unnecessary computational burden and accommodate real-time damage detection for time-varying systems, the previously discussed concepts of RPCA are restructured in conjunction with TVAR modeling.

Most physical systems function using data-based techniques that rely on the system's previous performance under similar loading patterns. For cases with nonstationary input excitation, it has been observed that RPCA-RRE methods discussed in Chapter 4, do not provide exact information about the damage instant. In such situations, system instabilities compounded with the nonstationary nature of the input mask the exact damage patterns, thereby making real-time spatio-temporal damage detection difficult (and in some cases, unattainable). Again, achieving exact detection results in real time for under determined systems with nonstationary input excitation data is an uphill task that necessitates certain modifications in the formulation of the FOEP based RPCA approach. Motivated by these practical performance issues, the basic RPCA based methodology can be tailor-made in order to improve the quality of real time detection. Towards this, TVAR modeling in conjunction with the existing RPCA approach is presented in this chapter. Central to the idea of basic time series models, the transformed response obtained from the RPCA algorithm is fitted through a TVAR model. This fundamental consideration is based on the belief that the transformed response through a preliminary whitening step will yield a model order much lower than the acquired physical response, which results in computationally effective analysis and facilitates a quick and reliable damage detection methodology. The resulting signal so obtained contains important information related to the spatial and temporal characteristics of damage, which are then examined in real-time to obtain the location and instant of damage in the system.

5.2 Background

Clearly, rapidly developing technologies will enhance the performance of an application-specific damage prognosis approach. Eventually, all monitoring strategies intend to perform in real-time (or near real-time) with an aim to provide cost-effective solutions, which are essentially non-invasive and have a dedicated sensing-processing-telemetry assembly for quick detection. Irrespective of the type of transducers, acquisition module and transmission cards used, damage detection fundamentally involves a case-specific algorithm (or a cohort of algorithms) for identifying structural damage patterns based on the output response acquired by the monitoring sensors (thereby accounting for an inverse dynamic problem). Methods based on decomposing a structural vibration response (such as principal component analysis (PCA) [1], frequency domain decomposition (FDD) [2], empirical mode decomposition (EMD) [3], to name a few) are highly sought after strategies that provide accurate representation of a possible damage instant and can be further utilized to eliminate outliers in case of noisy measurements. Research undertakings at both academic and corporate settings focus on developing robust, self-sustained, embedded sensing networks for a wide variety of engineering applications to identify structural damage in real-time [4–7]. In addition to statistical pattern recognition approaches to SHM, successful damage detection techniques function by powering the sensing systems that harvest the ambient energy available from the system's operating environment. Even though dense networks of sensors are typically desired for monitoring, their application must be related to the physics of the problem: the damage measurement assembly must embed *sensible* sensors that are responsive to the changes in structural parameters correlated with damage.

Since there will be a finite budget dedicated for sensing, data-driven models will often be a part of a study that will attempt to maximize the observability of damage given constraints on the sensing system parameters. As data becomes available at each instant of time, detection algorithms will be better equipped to deal with instabilities and nonstationarities that arise with the input excitation. Inevitably, algorithms must be designed in a way that addresses efficient real-time detection for short-paced measurements (acquired for a limited span) and must be able to accomplish stability in the early stages of data streaming. In this regard, RPCA will now be presented as a strong candidate towards real-time fault detection, damage pattern identification and localization – embedded within a single recursive framework that translates to low computational efforts.

The use of RPCA generates recursive eigenspace updates at each time stamp. This entails an automated online framework where TVAR modeling is shown to be applied in conjunction with DSFs for identifying the instant of damage [8, 10]. The use of an autoregressive (AR) process models the signal that is generated by one or more superimposed response – operating under the premise that past values obtained from a time series affect the current values – based on some statistical calculations [11, 15–17]. The integration of an AR model provides the added advantage of inferring the significant parameters of the underlying data, without resorting to spectral decomposition: this translates to the *first stage* of reducing the computational expense for the algorithm. The implementation of the AR algorithm towards structural damage detection is beyond the scope of this chapter and the readers are referred to seminal works [15, 18] on the topic for extensive details. An often-used approach to deploying a sensing system for damage detection – particularly associated with previous SHM studies [10, 19–22] – consists of using time series models to capture the key features of the acquired data series. The extracted features of interest are often selected in an ad-hoc manner without an apriori quantification of damage but are necessarily sensitive to the damage of interest. Similarly, traditional AR models track the transformed response for any deviations from the previously obtained values; however, these models are not amenable towards online implementation [10]. For real-time damage detection and analysis, AR models are tailored to better capture the non-stationarity involved with any data or due to the damage induced with the use of TVAR models. However, the paradigm of TVAR modeling can only be efficient if the signal possesses a relatively low model order (usually less than 5). As the inherent functioning of the RPCA algorithm is based on SVD of the response covariance matrix, the model order of the transformed response is automatically reduced (in some cases from 40 to about 3) in real-time. This enables a smooth functioning of the RPCA-TVAR module in an online mode and establishes the *second stage* towards computational reduction. As TVAR modeling is applied on the transformed responses (obtained from the RPCA algorithm) rather than the raw vibration responses, a, relatively low model order of the time-series model [9, 10] is sufficient to capture the dynamics of the structure in the transformed domain. In the proposed framework, damages in structures can be detected both spatially and temporally in real time.

5.3 RPCA and structural dynamics: A POC based formulation

While obtaining recursive eigenspace updates from structural response data, the eigen-vectors or proper orthogonal modes (POMs, discussed in detail in Chapter 4) generally obtain a biased significance. This importance – albeit appropriate – stems from the ability of the POMs to map themselves directly to the system damage, translating to a real-time capable monitoring module [10, 23]. Incidentally, eigenvalues generated during spectral decomposition go unnoticed except for tracking changes in magnitude [24] (where a reduction correlates to a damage) or for providing tuning information for tuned mass damper (TMDs) systems [25]. This chapter will focus on an eigenvalue *derivative* (not in a mathematical sense) in the form of POCs where the application of RPCA will be explained in the purview of structural dynamics.

In this context, a linear, classically damped, and lumped parameter system with mass, stiffness and damping matrices **M**, **C** and **K** subjected to an external force, with **x** as the displacement vector, is considered.

$$[\mathbf{M}]\{\ddot{x}(t)\}+[\mathbf{C}]\{\dot{x}(t)\}+[\mathbf{K}]\{x(t)\}=\{\mathbf{F}(t)\} \tag{5.1}$$

where $\mathbf{F}(t)$ is the input excitation which is assumed to be Gaussian and broadband. The symbols have their usual meanings as denoted in the previous chapter. The solution of the equation can be written as $\{\mathbf{x}\}_m = [\mathbf{V}]_{m\times s}\{\mathbf{q}\}_s$, where $\mathbf{x}$ is the measurement matrix of size $m \times N$, $\mathbf{q}$ is the matrix of corresponding modal coordinates of size $s \times N$. Here m represents the number of DOF and s corresponds to the number of modes considered with N sampling points. $[\mathbf{V}]_{m \times s}$ is the mode shape matrix which transforms the data from modal coordinates to the physical space. An important characteristics of the mode shape matrix is that each of its columns are orthogonal to each other with respect to the matrix $\mathbf{M}$. The covariance matrix of $\mathbf{X}$ ($\mathbf{R} = \frac{1}{N}\mathbf{X}\mathbf{X}^T$) can be expressed as $\mathbf{R} = \frac{1}{N}\mathbf{V}\mathbf{Q}\mathbf{Q}^\mathbf{T}\mathbf{V}^\mathbf{T}$.

From the above expression, $\mathbf{R_Q} = \frac{1}{N}\mathbf{Q}\mathbf{Q}^\mathbf{T}$ can be identified as the covariance matrix of the modal responses. For an undamped free vibration (i.e., $\mathbf{C} = 0$ and $\mathbf{F} = 0$ in Eq. 5.1), $\mathbf{R_Q}$ is exactly a *diagonal matrix*, while under mildly damped forcing conditions (i.e., $\mathbf{C} \neq 0$ and $\mathbf{F} \neq 0$ in Eq. 5.1), the matrix $\mathbf{R_Q}$ is an *approximately diagonal* matrix for a finitely large sample size. Under broadband excitations, the evolution of the physical response, $\mathbf{x}_i(t)$ and modal response $\mathbf{q}_i(t)$, in the time domain can be expressed in terms of the impulse response function as $q_i(t) = \int_0^\infty h(t-\tau) f_i(\tau)\, d\tau$, where $\mathbf{f}_i(\tau)$ represent the modal forces, related by the equation $\mathbf{f}_i(\tau) = v_i^T \mathbf{F}_i(\tau)$, where v_i represents the mode shape corresponding to the mode and $\mathbf{F}_i(\tau)$ represent the actual forces. The individual elements of the covariance matrix $\mathbf{R_Q}$ can be expressed as:

$$r_{ij}^Q = \int_{\tau=0}^{\infty} \int_{\theta=0}^{\infty} f_i(\tau) f_j(\theta) \left[\frac{1}{N} \sum_{i,j} h_i(t-\tau) h_j(t-\theta) \right] d\tau d\theta \tag{5.2}$$

Equation 5.2 shows that for a finitely large N, $\mathbf{R_Q}$ can be expected to be a diagonal matrix for an undamped system and nearly diagonal for a light to moderate modal damping. The POCs (ψ) or orthogonal transformation of the data can be written as a product of POMs ($\mathbf{W}$) and the data vector ($\mathbf{x}$) as:

$$\begin{aligned} \psi_i(t) &= \mathbf{W}^\mathbf{T} x_i(t) \\ &= q_i(t) + \gamma \end{aligned} \tag{5.3}$$

Hence POCs (ψ) can be expressed as a sum of true linear normal coordinates (q) and an error term (γ). To understand the behavior of the covariance matrix of the POC, $\mathbf{R}_\Psi = \frac{1}{N}\Psi\Psi^\mathbf{T}$ its essential to realize the individual elements of the $\mathbf{R}_\Psi$ matrix. Substituting from Eq. 5.3,

$$\begin{aligned} r_{ij}^{\psi} &= \frac{1}{N} \sum_{k=1}^{N} \psi_i(t_k) \psi_j(t_k) \\ &= \frac{1}{N} \sum_{k=1}^{N} [q_i(t_k) q_j(t_k) + \gamma q_j(t_k) + q_i(t_k)\gamma + \gamma^2] \end{aligned} \tag{5.4}$$

For practical systems having low to moderate damping and finite sample size, it can be understood from Eq. 5.4 that POC provide a good approximation to the true linear modal components which deviate from each other as damping increases [26,27]. Hence the POC covariance matrix $\mathbf{R}_\Psi$ is still expected to show a diagonally dominant behavior in the limit as $N \rightarrow \infty$ and when the errors are low (i.e., low to moderate damping) [10,26].

The POC matrix ,$\mathbf{\Psi}$ is obtained as:

$$\begin{aligned}\mathbf{\Psi} &= \mathbf{W}^{\mathbf{T}}\mathbf{X} \\ &= (\mathbf{V}+\varepsilon)^T\mathbf{X} \\ &= \mathbf{Q}+\mathbf{\Gamma}\end{aligned} \tag{5.5}$$

The main objective of RPCA is to provide recursive estimates of both $\mathbf{W}$ and ψ_k at each time instant. This integrates a sensing platform to a complete monitoring module as the analysis is carried out *in-situ*. For structural systems with low to moderate damping, obtaining the POCs at each time instant translates to obtaining the normal coordinates. Since the normal coordinates are independent in the modal domain, they are likely to be mono-component in nature, therefore, it can be conjectured at this stage that a lower model order will sufficiently explain the dynamics of the system. Moving on, the response covariance matrix $\mathbf{R}_k$ at any instant k can be written as a function of the covariance matrix of the previous time instant $\mathbf{R}_{k-1}$ and response vector $\mathbf{x_k}$ at the kth time instant as shown:

$$\mathbf{R}_k = \frac{k-1}{k}\mathbf{R}_{k-1} + \frac{1}{k}\mathbf{x}_k\mathbf{x}_k^T \tag{5.6}$$

Once the structure of the recursion is established, the covariance estimate R_k needs to be decomposed in the frequency domain. The spectral decomposition arises from the EVD of the matrix as: $[\mathbf{R}_k] = [\mathbf{W}_k]\,[\mathbf{\Upsilon}_k]\,[\mathbf{W}_k]^T$. On substituting in Eq. 5.6, the expression for the covariance matrix is rewritten as:

$$\mathbf{W}_k\mathbf{\Upsilon}_k\mathbf{W}_k^T = \frac{k-1}{k}\mathbf{W}_{k-1}\mathbf{\Upsilon}_{k-1}\mathbf{W}_{k-1}^T + \frac{1}{k}\mathbf{x}_k\mathbf{x}_k^T \tag{5.7}$$

The estimate of the POC vector at the kth time instant can be written as $\{\tilde{\psi}_k\} = [\mathbf{W}_{k-1}]^T\{\mathbf{x}_k\}$. On substituting in Eq. 5.7, the following expression can be obtained

$$[\mathbf{W}_k]\,[k\mathbf{\Upsilon}_k]\,[\mathbf{W}_k]^T = \mathbf{W}_{k-1}\left[(k-1)\,\mathbf{\Upsilon}_{k-1} + \tilde{\psi}_k\tilde{\psi}_k^T\right]\mathbf{W}_{k-1}^T \tag{5.8}$$

For the RPCA algorithm to be stable and robust, it is important that the term $[(k-1)\,\mathbf{\Upsilon}_{k-1} + \tilde{\psi}_k\tilde{\psi}_k^T]$ becomes *strongly diagonally dominant* (SDD). Based on system dynamics, it is observed that the use of Gershgorin's theorem [28,29] here provides an understanding of the eigenspace obtained through recursion. For a

SDD structure, the theorem states that the eigenvalues are close to the diagonal portion of the matrix, while the eigenvectors approximate to unity. Associated work by the authors [9, 10] have identified the term $\tilde{\psi}_k\tilde{\psi}_k^T$ to represent correlation between the POC estimates at a particular instant. Substituting from Eq. 5.3, the covariance between POC estimates can be written as:

$$\begin{aligned}\tilde{\psi}_k\tilde{\psi}_k^T &= \mathbf{W}_{k-1}^T x_k {x_k}^T \mathbf{W}_{k-1} \\ &= q_{k-1}{q_{k-1}}^T + \gamma {q_{k-1}}^T + \gamma^T q_{k-1} + \gamma\gamma^T\end{aligned} \tag{5.9}$$

As far as the dynamics of structural systems are considered, the error term in the Eq. 5.9 can be neglected with the increase in the number of sampling points. The magnitude of the error attenuates under moderate to low damping. The first term in Eq. 5.9 resembles the covariance of the normal coordinates at the instant $(k-1)$, which was found to be mono-component in nature. Hence, the term $q_{k-1}{q_{k-1}}^T$ represents a matrix whose diagonal terms dominate its off-diagonal terms; therefore, the term $\tilde{\psi}_k\tilde{\psi}_k^T$ can be safely assumed to be diagonally dominant. This in turn, ensures the diagonal dominance of the matrix $\left[(k-1)\Upsilon_{k-1} + \tilde{\psi}_k\tilde{\psi}_k^T\right]$, facilitating a rather straightforward application of Gershgorin's theorem – where it has been both applied and verified for structural systems. Since the terms involved are essentially rank-1 matrices subjected to one degree of perturbation, therefore, a first-order eigenperturbation (FOEP) is carried out which yields recursive eigenspace updates at each instant of time. Evidently, a less computationally intensive framework is automatically invoked since the ensuing FOEP works on the subsequent eigenspace updates and is not necessary for the covariance matrices at each stage of simulation. Therefore, the EVD of the matrix $\left[(k-1)\Upsilon_{k-1} + \tilde{\psi}_k\tilde{\psi}_k^T\right]$ can be substituted as $\mathrm{H_k}\boldsymbol{\Lambda}_k\mathbf{H}_k^T$ into Eq. 5.8 as,

$$[\mathbf{W}_k][k\Upsilon_k][\mathbf{W}_k]^T = [\mathbf{W}_{k-1}\mathbf{H}_k][\boldsymbol{\Lambda}_k][\mathbf{W}_{k-1}\mathbf{H}_k]^T \tag{5.10}$$

which yields the following iterative update equations:

$$\left.\begin{aligned}\mathbf{W}_k &= \mathbf{W}_{k-1}\mathrm{H}_k \\ \boldsymbol{\Upsilon}_k &= \tfrac{\boldsymbol{\Lambda}_k}{k}\end{aligned}\right\} \tag{5.11}$$

The recursive algorithm of Eq. 5.6 is transformed to obtain the values of H_k and Λ_k. Since the term $(k-1)\boldsymbol{\Upsilon}_{k-1} + \tilde{\psi}_k\tilde{\psi}_k^T$ is diagonally dominant, the eigen values can be assumed to be the diagonal entries of the matrix. Hence the ith diagonal entry of the term Λ_k can be represented as $(k-1)\lambda_i + \tilde{\psi}_i^2$, where λ_i is the (i,i) element of $\boldsymbol{\Upsilon}_{k-1}$ and ψ_i is the ith entry of the POC estimate. Once the eigen values are known, the corresponding eigen vectors can be found out, leading to H_k.

One of the key problems faced while applying FOEP approach is that the recursive eigen vectors obtained at each time instant are not ordered, which poses the problem of *permutation ambiguity* [10, 30]. This can be addressed by arranging the basis vectors according to a decreasing order of the corresponding eigenvalues in Υ_k. At each time instant, the eigenvalues indicate the contribution of the particular eigenvectors. The contribution factor can for a particular i^{th} eigen vector w_i at each time instant can be written as $\frac{\alpha_i^2}{\sum_{i=1}^{n} \alpha_i^2}$, where α_i^2 is the eigenvalue corresponding to w_i. Let $\mathbf{W}_k = [\mathbf{W}_k^1 \mathbf{W}_k^2]$, where $\mathbf{W}_k^1$ is the subspace at the kth time instant consisting of eigenvectors, that account for more than 90% of the energy in the participating modes of the system and $\mathbf{W}_k^1$ is the subspace at the kth time instant accounting for the remaining kinetic energy. These recursively estimated subspaces, are subsequently utilized to find the true POC at each instant of time as per the expression $\psi_k = \mathbf{W}_k^T \mathbf{x}_k$. The first element of the ψ_k vector represents the major principal component on which TVAR modeling is now performed.

5.4 TVAR modeling

The implementation of the RPCA algorithm ensures that the POCs are updated recursively. Once the transformed responses are obtained, the next logical step is to find out the instant of damage. In order to characterize the behavior of the POC updates (which can be approximated to normal coordinates), TVAR modeling is adopted here centred around the belief that time-varying coefficients are useful indicators of structural damage [10, 21]. Therefore, TVAR coefficients of the modeled transformed response are tracked in real time in order to identify the damage instant – either through the changes in the mean level of visualization or the presence of sudden spikes (crests or troughs) in the signal. The sudden changes in AR coefficients indicate the alterations in the dynamical properties of the system, such as shifts in natural frequencies and changes in mode shapes of the system induced due to the damage in the system. In the proposed work, the transformed response (i.e., the first POC) extracted from the RPCA algorithm is modeled using a TVAR model. The use of POCs instead of the raw vibration data – whose near resemblance to normal coordinates enables the use of a low model order – circumvents the issue of apriori model order selection.

From a mathematical perspective, let $\psi^1(k)$ represent the POC at any instant which captures the maximum kinetic energy of the system and let $v(k)$ denote the zero mean Gaussian white noise with variance σ_v^2. Then the AR model of order p can be represented as:

$$\psi^1(k) = \sum_{i=1}^{p} a_i \psi^1(k-i) + v(k) \tag{5.12}$$

At times when the input excitation is narrow-band, the eigenspace is correctly updated only if the POCs are considered nonstationary. This means that the AR coefficients automatically take the form of a time-varying vector – i.e., a_i becomes $a_i(t)$ – thereby allowing an easy apprehension of the system dynamics. For this purpose, the Kalman filter is utilized to estimate these time-varying coefficients, knowing the observations of the data [10, 31, 32]. The following equation is the discrete representation of the $a_i(t)$ coefficients and $w(k)$ is the process noise with variance σ_w^2 and covariance, $\mathbf{P_w} = \mathbf{I_{p\times p}}\sigma_\mathbf{w}^2$. Both noise measurements $v(k)$ and $w(k)$ are mutually independent and uncorrelated. Consider the following set of equations, where the unknown state vector $b(k)$ is expressed as shown:

$$\begin{aligned} \mathbf{b}(k) &= \mathbf{\Gamma}(k-1)\mathbf{b}(k-1) + \mathbf{w}(k) \\ \psi^1(k) &= \mathbf{C}(k)\mathbf{b}(k) + \mathbf{v}(k) \end{aligned} \tag{5.13}$$

The state vector $b(k)$ is given by: $\mathbf{b}(k) = \left[a_1(k), a_2(k), \ldots\ldots a_p(k)\right]^T$. The matrix $\mathbf{\Gamma}(k-1)$ is an identity matrix ($\mathbf{I}_{p\times p}$). The matrix C_k is the observation data with k discrete steps given as: $\mathbf{C}(k) = \left[\psi^1(k-1), \psi^1(k-2)\ldots\ldots\psi^1(k-p)\right]$. The Kalman filter has mainly two processes: one is the stepwise time update (prediction) and the other one is the measurement update (correction) of the predicted data. At each step, the set of the Kalman filter equations can be written as [10, 31, 32]:

$$\begin{aligned} &\mathbf{b}(k|k-1) = \mathbf{b}(k-1|k-1) \\ &\mathbf{P}_b(k|k-1) = \mathbf{P}_b(k-1|k-1) + \mathbf{I}\sigma_w^2 \\ &\psi^1(k|k-1) = \mathbf{C}(k)\mathbf{b}(k|k-1) \\ &\sigma_{\psi^1}^2(k|k-1) = \mathbf{C}(k)\mathbf{P}_b(k|k-1)\mathbf{C}(k)^T + \sigma_v^2 \end{aligned} \tag{5.14}$$

And,

$$\begin{aligned} &\mathbf{KG}(k) = \mathbf{P}_b(k|k-1)\mathbf{C}(k)^T\sigma_{\psi^1}^2(k|k-1)^{-1} \\ &\mathbf{b}(k|k) = \mathbf{b}(k|k-1) - \mathbf{KG}(k)\left[\psi^1(k) - \psi^1(k|k-1)\right] \\ &\mathbf{P}_b(k|k) = [\mathbf{I} - \mathbf{KG}(k)\mathbf{C}(k)]\mathbf{P}_b(k|k-1)] \end{aligned} \tag{5.15}$$

where $\mathbf{b}(k|k-1)$ represents apriori estimate and its linear combination would result in $\mathbf{b}(k|k)$, which is a posteriori. The Kalman gain $KG(k)$ gives a weightage to the prediction error $\psi^1(k) - \psi^1(k|k-1)$, to minimize the state estimation error $\mathbf{b}(k|k)$. This means that the apriori and the posteriori covariance estimates are given by $\mathbf{P}_b(k|k-1)$ and $\mathbf{P}_b(k|k-1)$, respectively. Although the use of TVAR models adequately capture the nonstationary nature of the transformed response, a stand-alone installation of such vectors in a monitoring system is not expected to provide useful visualization of the damage instant. Indicators that are amenable towards real-time implementation or damage sensitive features (DSFs) are then employed on the extracted TVAR feature space to identify both the instant and the location of structural damage. These statistical indicators function in real time and enable a smooth transition from a conventional monitoring integration to a real-time detection module.

5.5 Damage sensitive features

Tracking the eigenspace at each instant of time facilitates online processing of data streams acquired from each instrumented sensor. Irrespective of the damage inflicted to the system, the eigenspace by itself does not provide sufficient information regarding the spatio-temporal features of damage. DSFs play an important role in identifying the presence and location of damage (correct to the damaged DOF). The pursuit of a good DSF requires adequate performance in detecting the presence of damage, effectively distinguishing between the damaged and undamaged states of the structure, and the aptness to accurately locate and quantify the extent of damage. Additionally, to detect damage in real time, the chosen DSFs should be amenable towards online implementation and must be able to accurately identify the presence of damage with the progression of streaming data. Several damage detection and SHM techniques have been proposed in the literature [9, 13, 15, 17] that involve use of case-specific DSFs that indicate damage to the system through changes in their statistical descriptors. The use of traditional DSFs comes with a fair share of shortcomings such as the requirement of (**i**) baseline or reference data from a healthy structure; (**ii**) windowing of response data, (**iii**) remote detection analysis with impending transmission costs, and others which impede implementation in an online framework. The present portion of the chapter deals with a brief background on the formulation of *online DSFs*. For the purposes of this book, TVAR coefficients and recursive signal statistics on the acquired time-varying vectors are adopted for damage detection (in case of requirement of a new DSF, it will be separately introduced and studied in the respective chapter). The TVAR coefficients translate the *reorientation of eigenspace* due to damage to be manifested in the form of deviation in the mean level of the vector plots. Alternatively, the presence of a sudden spike accurately corresponds to the exact instant of damage in real-time. The following sections are dedicated to a brief understanding of the TVAR coefficients (a_i) and signal statistics on TVAR coefficients (μ_{a_i},ζ_{a_i}). For damage localization case studies, spatial RREs discussed in the previous chapter will remain the condition indicators for real-time damage localization.

5.5.1 Time varying auto-regressive coefficients

The estimation of AR coefficients frequently requires the use of windowing and baseline data to detect damage, making online implementation difficult. Motivated by these key shortcomings, a TVAR modeling based framework [10,

13] is implemented that is devoid of any baseline data and is essentially parameter-free. This enables the assembly to capture the nonstationary nature of the data associated due to damage. The proposed method tracks the TVAR coefficients online, which are used to ascertain the damage induced in the structure by a change in the mean level of the plot, at the exact instant of damage. The near mono-component nature of the transformed response ensures that low model order is sufficient to capture its dynamics; therefore, in the proposed framework, a model order of 2 (two) is pre-selected for all the cases. The basic Eq. 5.12 becomes:

$$\psi^1(k) = a_1(k)\psi^1(k-1) + a_2(k)\psi^1(k-2) + \nu(k) \tag{5.16}$$

where the symbols carry the same meaning as that of Eq. 5.12. Although a_1 and a_2 are expected to alter mildly at each time instants, the damage instant is characterized by sudden changes in the overall behavior of TVAR coefficients post damage. Tracking this change in a_1 and a_2 serves as an indicator of damage.

5.5.2 Recursive statistics on TVAR coefficients

The transformed response obtained after implementation of the RPCA algorithm sometimes shows modulatory/wavy behaviors, which can mask the accurate determination of damage instant. This is attributed to the non stationary and near mono-component nature of the transformed response ($\psi(t)$). As explained in Section 5.3, the POC tends to approximate normal coordinates only under certain assumptions which may not be realized in practice always. Hence, even though TVAR coefficients reflect the damage in a system, the change may not be always obvious. To address the aforementioned issue, recursive versions of the commonly used signal statistics are employed on TVAR coefficients [10]. The damage in a structural system is corroborated by the change in the behavior of TVAR coefficients which is manifested in the form of: (**i**) sharp peak in a_1 and/or a_2 in the vicinity of damage; (**ii**) drift of post damage a_1 and/or a_2 from the pre-damage values. This prompted the use of recursive mean (μ_{a_i}), and higher moments (ζ_{a_i}) of AR coefficients which are expected to capture the essence of damage better and validate the findings obtained from the TVAR coefficients.

$$\begin{aligned}
\mu_{a_i}(k) &= \frac{k-1}{k}\mu_{a_i}(k-1) + \frac{1}{k}a_i(k) \\
\sigma^2{}_{a_i}(k) &= \frac{k-1}{k}\sigma^2{}_{a_i}(k-1) + \frac{1}{k}a_i(k)a_i(k) \\
\zeta_{a_i}(k) &= \frac{(k-1)\zeta_{a_i}(k)/k + a_i(k)a_i(k)a_i(k)a_i(k)a_i(k)a_i(k)/k}{\left((k-1)\sigma^2_{a_i}(k-1)/k + a_i(k)a_i(k)/k\right)^3}
\end{aligned} \tag{5.17}$$

where ζ_{a_i} refers to the sixth order moment and $\sigma^2_{a_i}$ represents the second moment or variance.

5.6 Proposed algorithm

The overall methodology followed in this section entails the integration of separate modules which detect two key ingredients – *temporal* and *spatial* damage detection, –simultaneously. The first module deals with the global damage detection, wherein the raw acceleration data is processed by the RPCA algorithm as the data streams in real time. This accounts for an online damage detection framework as no batches of data are utilized in order to form a baseline. TVAR modeling is carried out on the updated first principal component which yields TVAR coefficients at each instant of time. As previously explained, TVAR coefficients are tracked online for any major and minor changes. DSFs (like higher moments) are utilized to corroborate the instant of damage, that provides a validation to the plots obtained from the TVAR coefficients. Once the damage instant is detected, the spatial damage detection module further resolves the location of damage.

For easy comprehension, the basic steps of the algorithm are enumerated as follows:

1. First, batch PCA is employed on some initial data points in order to estimate the initial eigen vector and eigen value matrices, i.e., the initial eigenspace.

2. The RPCA algorithm then operates online on the real time input of the streaming data.

3. Using the recursive gain depth parameter, the covariance estimate at the present time instant is derived using the covariance estimate at the preceding time instant. From the recursive updates, the eigen vector and eigen value matrices are updated using the FOEP approach and the transformed responses (principal components) are obtained using the RPCA algorithm.

4. Suitable time series models are generated based on the responses to fit a TVAR model. The DSFs are tracked in real time in order to extract the changes in the model coefficients, thereby revealing the faults in the system.

5. Once the instant of damage is determined, the algorithm shifts on to the next module where the spatial detection of damage takes place. The local RREs are tracked online recursively to capture the spatial effect of the damage, visually.

Vibration responses are processed by the RPCA algorithm to obtain the transformed responses. Corresponding to these lower order near mono-component responses, TVAR modeling is utilized to extract the time-varying coefficients through which damage instant is detected. It should be mentioned here that the proposed RPCA-TVAR has the following few characteristics: (**i**) the data is processed at each time instant, as and when it becomes available – i.e., the algorithm works *online*. (**ii**) to locate the instant of damage a reference value (baseline) is not required, i.e., it is *baseline free*. (**iii**) there are no parameters controlling the working of the algorithm, hence its *parameter free*. The above mentioned characteristics of the proposed algorithm makes it an ideal choice for a real time damage detection framework.

5.7 Numerical example

The B-W model described in the previous chapter (Section 4.6.1) is taken into consideration for damage detection studies. The reason for choosing the same model for numerical case studies is to identify the improvement in the detection prowess of the newly proposed RPCA-TVAR algorithm. The parameters of the model remain same as previously discussed, and assume a linear dependence on DI with a change in the then nonlinear force term. The B-W system is chosen for an intuitive understanding of the behavior of the RPCA-TVAR algorithm towards validating the spatio-temporal aspect of damage in real time. The current study considers *global damage* through a change in the nonlinear force term κ, while the concept of *local damage* is emulated through a reduction in the linear storey stiffness. The detection studies are presented for the system subject to white noise excitation followed by certain key results and observations.

5.8 Temporal damage detection results

Typical recorded accelerations at the roof of the structure subjected to various levels of nonlinearity do not provide sufficient information regarding the presence of damage (if at all, any). In order to identify the damage occurred, the raw vibration responses need to be processed with the newly proposed RPCA-TVAR algorithm. The damage is induced through change of nonlinearity by 15% and 30% at a particular time instant of 31 s. In order to detect the temporal damage, two DSFs are used: TVAR coefficients and their higher order moments (HOMs). It can be clearly observed from Fig. 5.1 that the damage instant can be easily detected using the TVAR coefficients by exploiting their sudden changes in their mean level at the damage instant. Thus, it confirms that TVAR modeling is amenable to online damage detection in a recursive framework. Since the transformed responses obtained using the RPCA algorithm are mono-component in nature, a relatively low model order (say, 2) can be used, which is shown in Fig. 5.1. Additionally, it is well understood from Fig. 5.1(b) that the recursive mean could capture the essence of fault detection where it shows a significant deviation in its original trajectory at $31s$. However, it will be shown in the following sections that the plots for HOMs provide more accurate results than the recursive mean as observed for the experimental case.

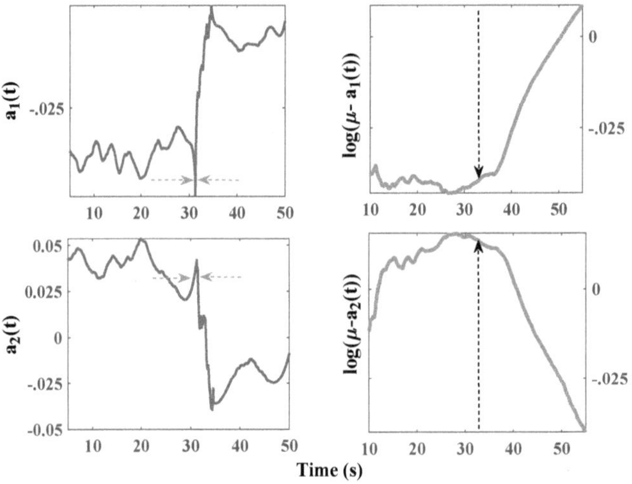

Figure 5.1: Damage detection using damage sensitive features for 30% non linearity change.

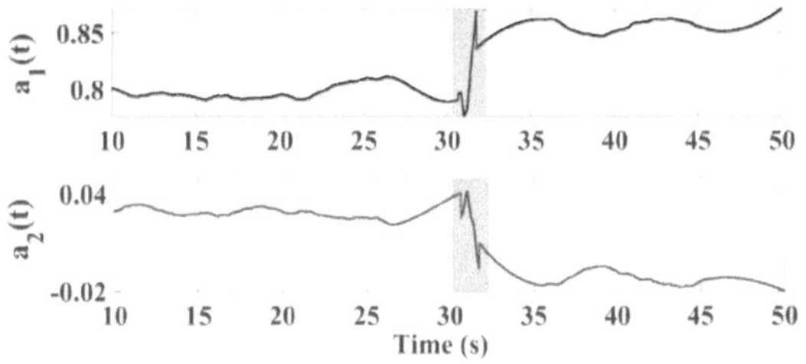

Figure 5.2: Damage detection using TVAR coefficients for 15% non linearity change.

The efficacy of recursive DSF is less effective for lower percentage changes in nonlinearity. As evident from literature, damages of the order of 25% have been often reported as a lower limit for vibration based damage detection [8]. However, in the present study the use of TVAR coefficients successfully detects damage corresponding to a 15% change in nonlinearity. As seen from Fig. 5.2, the TVAR plots serve as a robust DSF for detecting damage. The damage instant can be clearly identified from the figure as $31s$ using the proposed method.

In order to further validate the damage instant obtained from the TVAR plots, the use of recursive mean on the TVAR coefficients is tracked to show any significant changes, online. By looking at Fig. 5.3(a) and 5.3(b), it can be clearly observed that there is a significant change in the estimate of the recursive mean. It can also be inferred from the figures that the DSF utilized for tracking the changes shows distinct peaks indicating the exact instant of the damage that has occurred

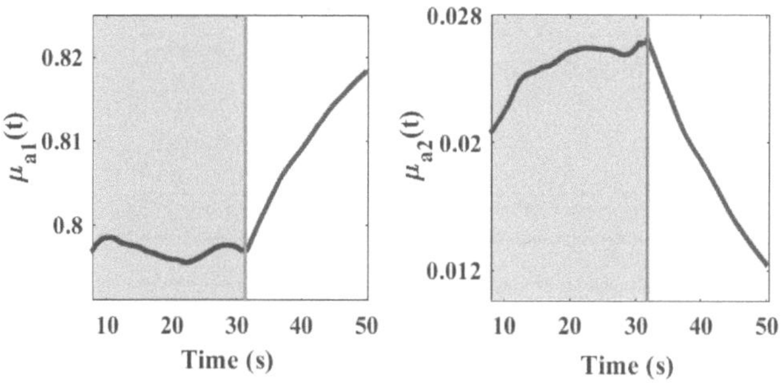

Figure 5.3: Damage detection using recursive mean for 15% non linearity change.

in the system. It becomes imperative to review and report the computational efficiency of the proposed method. As previously established, damage being a real time event necessitates that the algorithm assigned to track it in real time must function online, with less computational exhaustion and provide detection results as and when the vibration data streams in. In this context, the time taken for a single iteration of the algorithm is found to be 4.59 milliseconds, which closely emulates a real time process. The arrangement of basis vectors in the decreasing order of the corresponding eigenvalues consumes 79.2 microseconds, which is 1.7% of the total time consumed for a complete single iteration. However, the time consumed depends upon the computational power of the system used and is more efficient for systems with superior processing power, higher gigabytes of RAM, multiple cores, and more.

5.8.1 Spatial damage detection results

In the previous sections, the notion of global damage had been discussed through variation in the nonlinear force term κ right in the differential equation of motion of the system. In the present section, the local damage to the structure is perceived to be an alteration in the linear storey stiffness for a particular floor. Together, the pair of individual damage patterns subsequently correspond to *online spatio-temporal* damage detection. As already mentioned, only when the instant of damage in ascertained, the algorithm shifts on to the spatial damage detection module in order to exactly identify the location of the damage. The simultaneous temporal and spatial case deals with a 35% change in linear stiffness in the 3rd storey of the structure. The system is assumed to be fully nonlinear by scaling the value κ to 1. Once the damage instant is detected (by tracking the change in the mean of the TVAR coefficients), the spatial module starts functioning online in order to localize the position of the damage that has occurred in the system. From Fig. 5.4, it can be shown that the clear damage instant is identified at 31s

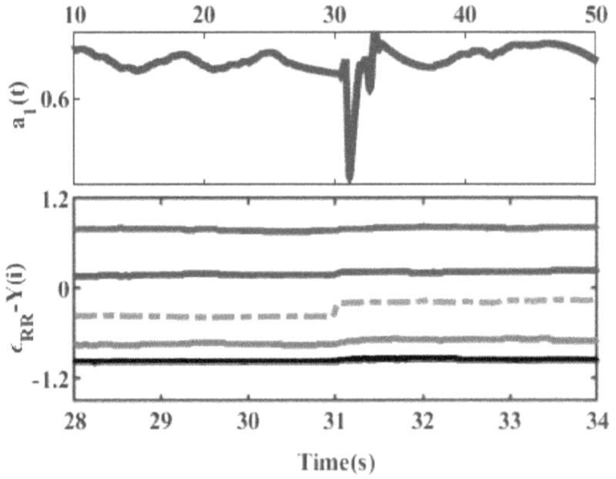

Figure 5.4: Comparison between spatial and temporal damage for a 35% change.

for a 35% damage. Once the damage instant is detected, the spatial RRE in the neighborhood of the vicinity of damage (say, 28s to 34s) is examined. From the same figure, it is clearly observed that the spatial RRE for a third degree of freedom shows a significant change at 31s compared to the other set of responses, thereby indicating that the damage has occurred in the third storey. As outlined before, the temporal and spatial detection module should run simultaneously. The changes in the TVAR coefficients and the local RRE are tracked online recursively in order to capture the essence of temporal and spatial damage detection, respectively. However, it is worth mentioning the fact that as the extent of stiffness degradation decreases, detecting spatial damage becomes increasingly difficult as compared to detecting temporal damage *alone*.

For the case of local damage, the nonlinear force term is kept constant at unity, which indicates that the value of the damage index (DI) is zero, in this case. As far as the system dynamics is concerned, the damage is manifested in the form of a linear stiffness change, identified as local damage, which is captured by the temporal damage detection module and is reflected by a peak in the TVAR plot, shown in Fig. 5.4. The phenomenon of local damage, unlike the change in nonlinearity at the base of the model, is confined only to a particular DOF and does not affect the structure as a whole, which is clearly observed from the plot of spatial RREs in Fig. 5.4(b). The plot of the TVAR coefficients retains its original trajectory even after the algorithm indicates the damage instant not because the system is repaired but rather the strength of frequency burst ceases to maintain its localization and intensity after a certain point of time.

The use of RREs as an effective damage localization tool emanates from the idea that spatial RREs show significant distortions in the trajectory only in the DOF that has undergone damage in the structure. It is further assumed that

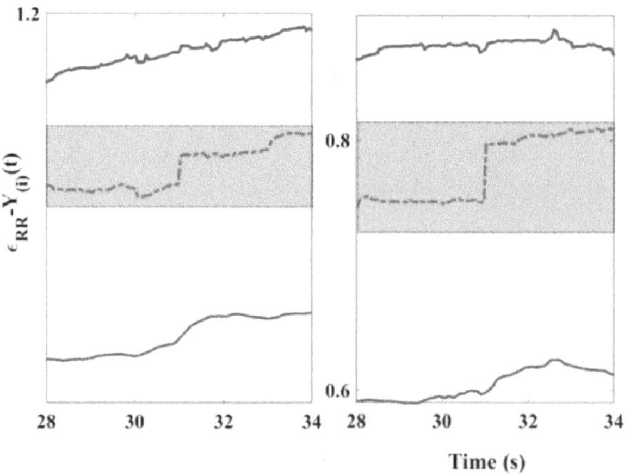

Figure 5.5: Spatial damage detection for negligible non-linearity.

the presence of nonlinear force at the base is unlikely to propagate significantly to the individual DOFs, which is quite valid, considering the weakly nonlinear nature of the system. To substantiate the point, a simple case is implemented to demonstrate that the nonlinearity associated at the base doesn't have a significant effect in reducing the efficacy of RRE. The damage is simulated at the local level by reducing 3rd storey linear stiffness by 25% and 35%, respectively, at a particular time instant of $t = 31$ s for the case of κ=0.02 (nonlinear force nearly equal to zero) and compared with the aforementioned κ=1 (hysteretic nonlinear force is fully present) case. In both the cases, $\varepsilon_{RR} - Y_3$ is able to detect local damage quite clearly as evident from the Fig. 5.5 (and Fig. 5.4). This clearly shows that the presence of hysteretic nonlinearity at the base doesn't affect the real time detection of damage significantly at least for the above case. However, the proposed methodology might not be suitable if the individual DOFs are modeled as nonlinear oscillators as the characterization will be stronger and more prominent.

5.8.2 Results for El Centro ground excitation

To illustrate the potential application of the proposed method, a numerical simulation is performed on the B-W system excited using El Centro recorded vibration data. The damage is simulated for a 30% change at 25s. The proposed algorithm processes the data and simultaneous tracking of the DSFs are carried out recursively. From Fig. 5.6, it can be observed that the plots of TVAR coefficients do not indicate a clear instant of damage due to the highly nonstationary behavior of the input excitation. In order to alleviate this shortcoming (rather, mask the non-

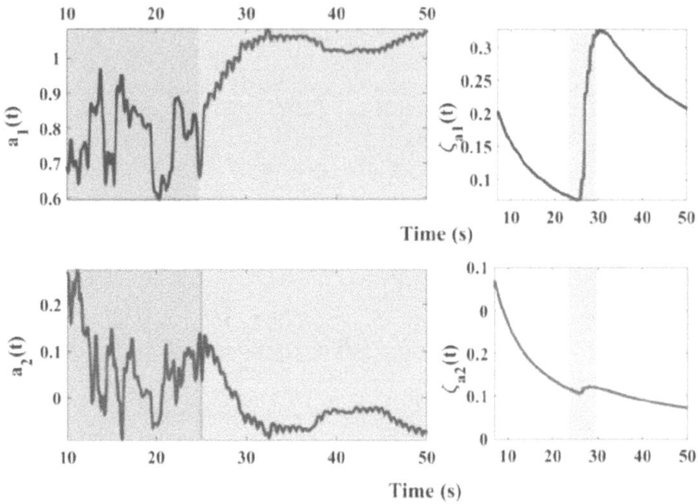

Figure 5.6: DSF for El Centro excitation for 30% change.

stationary attribute) of the data, HOMs of the TVAR coefficients are employed to clearly indicate the instant of damage. The effect of structural damage on the HOM plots is much more significant than the effect of local non-stationarity associated with the input excitation. This makes the HOMs (ζ_{a_i}) better equipped to capture the essence of damage. In the present damage scenario, the sixth moment of the TVAR coefficients($\zeta_{a_1}(t)$, $\zeta_{a_2}(t)$) is effectively used to indicate the exact instant of damage. Apparent from Fig. 5.6, plots for sixth order cumulant for $a_1(t)$ and $a_2(t)$ show a sudden change in the mean level. While approaching the damage instant, the plots show a significant change in the mean level which validates the occurrence of damage at t = 25s.

5.8.3 Results for time-diluted damage

This section mainly deals with the effect of a time diluted nonlinearity change due to progressive degradation of the base isolator. Progression of degradation is considered at a rate of 2% nonlinearity change (κ) every 2s. This phenomena is closely related to real life cases where a structure deteriorates due to the aftermath of an earthquake and at each passing second, the system undergoes a transitory state of collapse. Deterioration can be directly attributed to the loss of stiffness, a numerical study of which is carried out in this section and the results are reported for inference. In this context, simulation studies are performed where the value of κ changes at a steady rate for a period of 50s. It can be clearly observed from Fig. 5.7(a) and Fig. 5.7(b) that the AR coefficients continuously show a change in the

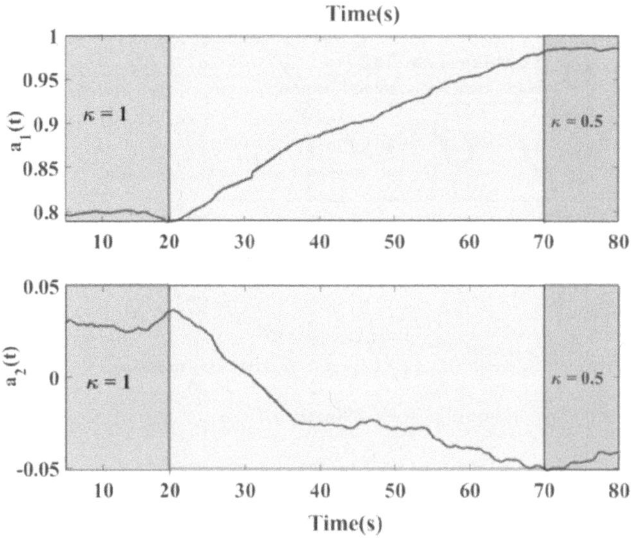

Figure 5.7: Damage detection using AR coefficients for a time-diluted damage.

mean level from 20 seconds (which indicates the commencement of damage) until 70 seconds, when the time-diluted damage has ended over a period of 50 seconds. This ultimately results in a change in the value of the nonlinear force term κ from 1 to 0.5, indicating the period of slow time-diluted degradation induced in the system. This simulated event is successfully captured by the TVAR coefficients. The results obtained from Fig. 5.7 clearly indicate the efficacy and the robustness of the algorithm to capture a *time-diluted progressive damage*. It can be clearly observed from Fig. 5.7(a) and Fig. 5.7(b) that the TVAR coefficients continuously show a change in the mean level from 20s (which indicates the commencement of damage) until 70 seconds, when the time-diluted damage has spanned over a period of 50 seconds. This results in an alteration in the value of the nonlinear force term κ from 1 to 0.5, which indicates the period of slow time-diluted damage induced in the system. This degradation event is successfully detected by the RPCA algorithm by tracking the TVAR coefficients over a timespan of 50s.

5.8.4 Results for underdetermined case-white noise excitation

To assess the applicability of the proposed method for a more practical situation, consider again an under determined system, similar to the one described in the previous chapter. From a practical viewpoint, such a system arises in flexible structures that are instrumented with a relatively smaller number of sensors because of cost and other factors. The proposed algorithm tries to address online

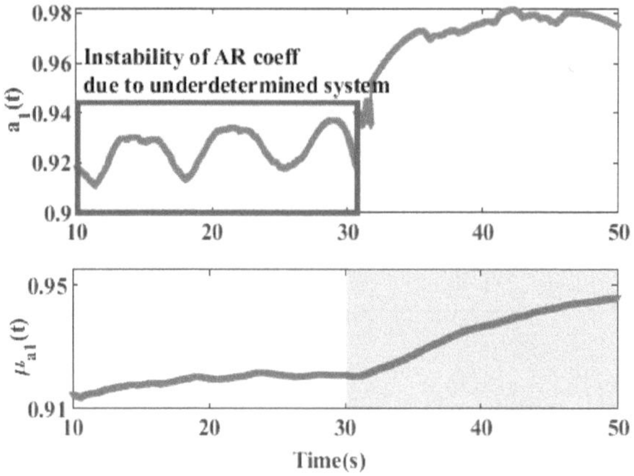

Figure 5.8: DSF for under determined case 1-Global Damage, 20% change in nonlinearity.

damage detection by assuming the number of (instrumented) DOF in the structure to be equal to number of available measurements. Since in the proposed algorithm, TVAR modeling is carried out on the first transformed response, the detection finesse is not compromised for an under determined system. Theoretically, the number of DOFs instrumented should be at least equal to number of actively participating modes (a basic concept of spectral decomposition, [8, 10]). The vibration response data from the top few DOFs is of foremost importance as it has the maximum effect on the first modal response and retains the major information not only on the occurrence of damage but also on the system kinetics. As previously discussed for the RPCA-RRE approach, the B-W system is studied, where the recorded acceleration response data is considered to be unavailable from the second and fourth floors. Damage was induced for the following two cases (i) Sudden change in nonlinear base isolator force at $31s$ (ii) Local damage induced in the 3rd storey at $31s$.

Considering the first case, the damage is induced through a 20% change in nonlinearity at a particular instant. From Fig. 5.8, it is clear that the algorithm is able to perceive damage using TVAR coefficients even without the availability of the complete set of responses. It is worth noting the fact that the number of DOF selected will have an impact on the detectability limit. For the above case, the algorithm was found to detect changes in nonlinearity up to 20%. On the other hand, for the second case, a 30% reduction in stiffness was induced in the fourth storey column at 31s and vibration responses from all the floors except the second and fourth were made available as input to the algorithm. The results are as shown in Fig. 5.9. From both the figures it can be observed that there is a lot of activity present before the instant of damage is attained. This

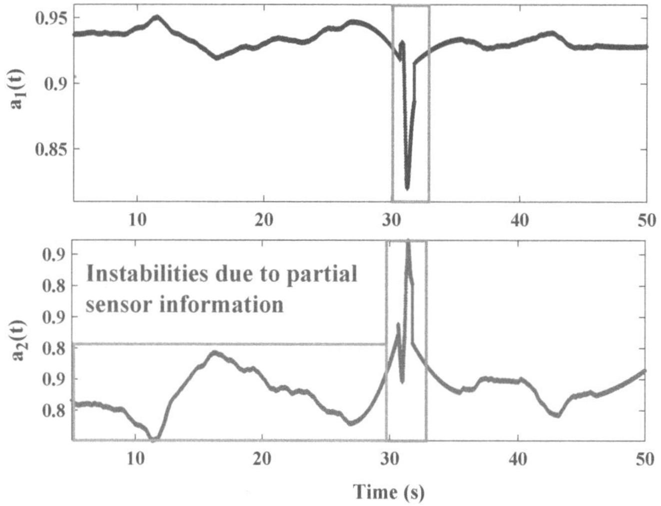

Figure 5.9: DSF for under determined case 2- Local Damage, 30% change in stiffness.

can be purely attributed to the instabilities that arise in the structuring of the algorithm due to unavailability of a complete set of sensor responses. Albeit these instabilities, it can be safely concluded that the algorithm is well equipped to estimate spatio-temporal damage even for under determined systems, which is a key entitlement of the present chapter. From the results obtained, it can be clearly understood that the applicability of the proposed method towards dealing with nonstationary excitations for nonlinear systems outperforms the detection potential of the RPCA-RRE based scheme.

5.9 Experimental study

The aluminum beam experimental setup described in the previous chapter is considered for carrying out damage detection studies using the proposed RPCA-TVAR based approach. The use of the same setup is justified from a utility perspective – primarily due to the fact that the proposed methodology provides a better set of detection results as compared to the ones obtained from the RPCA-RRE scheme – especially if a nonstationary input excitation is anticipated (in the event of earthquakes). As the experimental setup was provided in the previous chapter, the same is skipped here for brevity. Abiding by the prior experimental scheme, the trials are carried out by subjecting the aluminum beam to a scaled ground motion (1999 Chi-Chi ground motion, scaled to peak $0.3g$). In order to simulate a real time damage scenario, the rubber strip attached to the free end of the cantilever beam is snapped accurately at a fixed time instant, during its

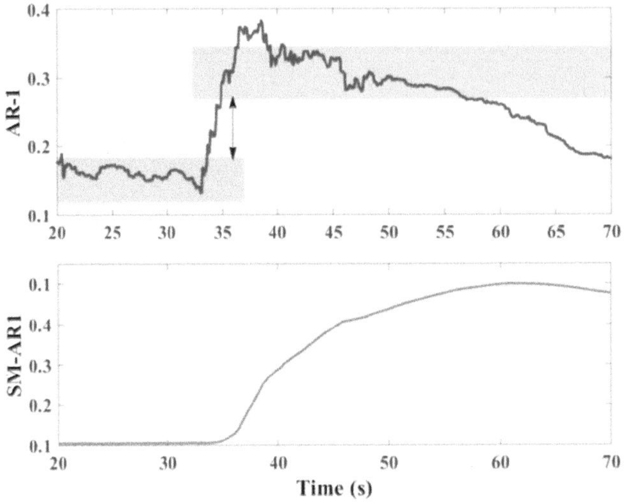

Figure 5.10: DSFs used for an experimental trial.

shaking motion. The recorded output acceleration plots obtained from the sensors do not provide a visual cue as to when the damage has occurred.

On the contrary, the instant of damage is well reflected by the plot of the DSF, which cannot be directly perceived from the raw acceleration data recorded from the sensors. It is duly noted from Fig. 5.10 that the TVAR coefficient $a_1(t)$ shows a significant change in mean at 33s (indicated by the double-arrow), that corresponds to the exact instant of damage for the experimental trial. Accompanying this figure, is the plot of the the HOM ($\zeta_{a_i}(t)$) that indicates a clear change in the mean level at 33s. This verifies the accurate instant of the rubber-snap, an event that closely corresponds to a real life damage. Hence, it can be concluded that the proposed method provides quality detection results even when the nature of the excitation is predominantly, nonstationary.

References

[1] Tibaduiza, D. A., Mujica, L. E. and Rodellar, J. (2013). Damage classification in structural health monitoring using principal component analysis and self-organizing maps. Structural Control and Health Monitoring, 20(10): 1303–1316.

[2] Brincker, R., Andersen, P. and Cantieni, R. (2001). Identification and level I damage detection of the Z24 highway bridge. Experimental Techniques, 25(6): 51–57.

[3] Xu, Y. L. and Chen, J. (2004). Structural damage detection using empirical mode decomposition: experimental investigation. Journal of Engineering Mechanics, 130(11): 1279–1288.

[4] Light, G. M., Kwun, H., Kim, S. Y. and Spinks Jr, R. L. (2002). U.S. Patent No. 6,396,262. Washington, DC: U.S. Patent and Trademark Office.

[5] Giurgiutiu, V. and Xu, B. (2007). U.S. Patent No. 7,174,255. Washington, DC: U.S. Patent and Trademark Office.

[6] Brownjohn, J. M. (2007). Structural health monitoring of civil infrastructure. Philosophical Transactions of the Royal Society A: Mathematical, Physical and Engineering Sciences, 365(1851): 589–622.

[7] Glisic, B. and Inaudi, D. (2008). Fibre Optic Methods for Structural Health Monitoring. John Wiley & Sons.

[8] Krishnan, M., Bhowmik, B., Tiwari, A. K. and Hazra, B. (2017). Online damage detection using recursive principal component analysis and recursive condition indicators. Smart Materials and Structures, 26(8): 085017.

[9] Bhowmik, B., Tripura, T., Hazra, B. and Pakrashi, V. (2019). First-order eigen-perturbation techniques for real-time damage detection of vibrating systems: Theory and applications. Applied Mechanics Reviews, 71(6).

[10] Krishnan, M., Bhowmik, B., Hazra, B. and Pakrashi, V. (2018). Real time damage detection using recursive principal components and time varying auto-regressive modeling. Mechanical Systems and Signal Processing, 101: 549–574.

[11] Bhowmik, B., Krishnan, M., Hazra, B. and Pakrashi, V. (2019). Real-time unified single-and multi-channel structural damage detection using recursive singular spectrum analysis. Structural Health Monitoring, 18(2): 563–589.

[12] Bhowmik, B., Tripura, T., Hazra, B. and Pakrashi, V. (2020). Real time structural modal identification using recursive canonical correlation analysis and application towards online structural damage detection. Journal of Sound and Vibration, 468: 115101.

[13] Bhowmik, B., Tripura, T., Hazra, B. and Pakrashi, V. (2020). Robust linear and nonlinear structural damage detection using recursive canonical correlation analysis. Mechanical Systems and Signal Processing, 136: 106499.

[14] Kijewski-Correa, T., Kwon, D. K., Kareem, A., Bentz, A., Guo, Y. et al. (2013). SmartSync: An integrated real-time structural health monitoring and structural identification system for tall buildings. Journal of Structural Engineering, 139(10): 1675–1687.

[15] Farrar, C. R. and Worden, K. (2007). An introduction to structural health monitoring. Philosophical Transactions of the Royal Society A: Mathematical, Physical and Engineering Sciences, 365(1851): 303–315.

[16] Balageas, D., Fritzen, C. P. and Güemes, A. (eds.). (2010). Structural Health Monitoring (Vol. 90). John Wiley & Sons.

[17] Farrar, C. R. and Worden, K. (2012). Structural Health Monitoring: A Machine Learning Perspective. John Wiley & Sons.

[18] Lakshmi, K., Rao, A. and Gopalakrishnan, N. (2016). Singular spectrum analysis combined with ARMAX model for structural damage detection. Structural Control and Health Monitoring, Wiley Online Library, 24(9).

[19] Krishnan Nair, K. and Kiremidjian, A. S. (2007). Time series based structural damage detection algorithm using Gaussian mixtures modeling.

[20] Nair, K. K., Kiremidjian, A. S. and Law, K. H. (2006). Time series-based damage detection and localization algorithm with application to the ASCE benchmark structure. Journal of Sound and Vibration, 291(1-2): 349–368.

[21] Musafere, F., Sadhu, A. and Liu, K. (2015). Towards damage detection using blind source separation integrated with time-varying auto-regressive modeling. Smart Materials and Structures, IOP Publishing, 25(1) 015013.

[22] Hazra, B., Sadhu, A., Roffel, A. J. and Narasimhan, S. (2012). Hybrid time-frequency blind source separation towards ambient system identification of structures. Computer-Aided Civil and Infrastructure Engineering, 27(5): 314–332.

[23] Hearn, G. and Testa, R. B. (1991). Modal analysis for damage detection in structures. Journal of Structural Engineering, 117(10): 3042–3063.

[24] Kim, J. T., Ryu, Y. S., Cho, H. M. and Stubbs, N. (2003). Damage identification in beam-type structures: Frequency-based method vs mode-shape-based method. Engineering Structures, 25(1): 57–67.

[25] Verdirame, J. and Nayfeh, S. (2003, April). Design of multi-degree-of-freedom tuned-mass dampers based on eigenvalue perturbation. In 44th AIAA/ASME/ASCE/AHS/ASC Structures, Structural Dynamics, and Materials Conference, p. 1686.

[26] Feeny, B. (2002). On proper orthogonal co-ordinates as indicators of modal activity. Journal of Sound and Vibration, Elsevier, 255(5): 805–817.

[27] Feeny, B. F. and Kappagantu, R. (1998). On the physical interpretation of proper orthogonal modes in vibrations. Journal of Sound and Vibration, Elsevier, 211(4): 607–616.

[28] Golub, G. H. and Zha, H. (1994). Perturbation analysis of the canonical correlations of matrix pairs. Linear Algebra and its Applications, 210: 3–28.

[29] Bell, H. E. (1965). Gershgorin's theorem and the zeros of polynomials. The American Mathematical Monthly, 72(3): 292–295.

[30] Sadhu, A. and Hazra, B. (2013). A novel damage detection algorithm using time series analysis-based blind source separation. Shock and Vibration, IOS Press, 20(3): 423–438.

[31] Yang, J. N., Lin, S., Huang, H. and Zhou, L. (2006). An adaptive extended Kalman filter for structural damage identification. Structural Control and Health Monitoring, Wiley Online Library, 13(4): 849–867.

[32] Chatzi, E. N. and Smyth, A. W. (2009). The unscented Kalman filter and particle filter methods for nonlinear structural system identification with non-collocated heterogeneous sensing. Structural Control and Health Monitoring, Wiley Online Library, 16(1): 99–123.

Chapter 6

Multi-sensor Real Time Damage Detection Techniques (B): Real Time Structural Damage Detection Using Recursive Canonical Correlation Analysis

In this chapter, a new real time structural damage detection strategy is developed that aims at identifying spatio-temporal damage patterns, online. The method, known as *recursive canonical correlation analysis* (RCCA), is a direct recursive extension of the traditional *canonical correlation analysis* (CCA). The chapter begins with a brief introduction to the need for using the algorithm and moves on to the detailed methodology developed for detecting damage in real time. Numerical simulations carried out on both weakly and strongly nonlinear systems demonstrate the applicability of the method and its exclusive superlative performance over the previously discussed RPCA algorithm, that greatly motivates the scope for implementing the proposed method. Experimental verifications on an idealized 2 DOF shear building model, followed by a full-scale demonstration of the method on the UCLAFB, demonstrate the robustness of the algorithm. The key results and discussions are presented next, followed by some important conclusions. The important acronyms used in this chapter are provided in Table 6.1.

Table 6.1: Important acronyms.

CCA	Canonical Correlation Analysis
RCCA	Recursive Canonical Correlation Analysis
ICA	Independent Component Analysis

6.1 Motivation

A majority of the response based detection schemes comprise of data-driven statistical techniques developed primarily in the field of classical time series analysis, signal processing algorithms and multi-variate statistics, that show enormous potential in identifying the key patterns of these series. In this context, algorithms premised on EVD of the covariance matrix of the physical response, such as PCA [1], have enormously contributed to a multitude of structural dynamics applications. A powerful statistical technique developed in the recent years, CCA, utilizes the block covariance matrix of the physical responses in order to measure the underlying correlation between two sets of multidimensional variables [2–13]. The primary objective of CCA is to find projections such that the mutual correlation coefficient is maximized [11–15]. In addition to multivariate statistical analysis [6, 7], statistical imaging [9, 10], social and behavioral depiction studies [4], the underlying concepts of CCA are utilized in finding key patterns of time series [13] for both linear and nonlinear systems [14], and are also applied in meteorological studies [8] and flood frequency estimation [12]. The method takes into account the eigenspace characteristics, which could be globally tracked to ascertain the damage occurred to the system [15]. The plethora of SHM literature provides ample instances where CCA based applications involving BSS algorithms have been implemented in the context of modal identification of structures subjected to ambient vibrations [16, 17]. CCA promises to incorporate a mathematically general structure consistent with the theoretical advancements in structural dynamics that can possibly accommodate both damage detection and modal identification in a single framework. To date, the attempt at using CCA in its unabridged version at identifying damage patterns in real time, is a novel work undertaken in this research. Again, addressing the facets of real time modal identification still poses a big challenge, owing to permutation ambiguity. The aspect of unifying CCA to address both modal identification and damage detection in real time in a simultaneous framework poses a formidable challenge that is beyond the scope of this research and is kept as a possible extension to be dealt with in the future.

The utility of CCA as a structural damage detection tool can be envisaged from the successful implementation of BSS towards modal identification [16, 17] and damage detection [18, 19], in the recent times. Techniques involving separation of modes, such as BSS and ICA, emerge as potential candidates for modal identification algorithms. Since a class of BSS methods come under the category of CCA based schemes [21], direct application of CCA towards modal identification and damage detection is imperative. However, it is worth noting the fact that the majority of the available damage detection schemes are *offline*, requiring windowing of the data in order to compare the newer set of responses to the baseline values [20]. An indispensable requisite for any detection algorithm to identify fine levels of damage, is its ability in comparing the recorded data at continuous intervals, from which the damage to the structure can be detected. This necessitates that the algorithms must work *online*, an aspect that has rarely been reported in the context of damage detection literature. As damage is a real time event, it further warrants that the detection algorithm must work online, which motivates the utilization of RCCA, a direct extension of the traditional CCA approach using the previously slated FOEP theory [22, 23]. In line with the previous salient theoretical treatment of the topic, the FOEP technique provides the rank-one eigenvector updates at each instant, as and when the vibration data streams in real time. The use of RCCA as an effective damage detection tool emanates from the conceptualization of the distortion of the eigenspace, that can be modeled using TVAR and utilized to identify the spatio-temporal damage in a simultaneous framework.

The implementation of the recently established method, RPCA, is based on the principles of orthogonal transformation of the data covariance matrix arising from an uniaxial vibration. The CCA method, however, generates a block covariance matrix utilizing multi-directional data, that arises mostly during field studies and experimental setups. It is observed that the RPCA based methodologies consistently fail to perform using multi-directional data, as opposed to the CCA based methods, where the inherent formulation of the technique pivots around this concept. It can therefore, be surmised, that the RCCA method should provide better detectability results for multi-directional dependencies and flexible structures with torsionally coupled systems as compared to the previously discussed RPCA scheme, which motivates the present study. Due to its more robust and accommodative essence, the RCCA method can be applied for a diversified range of excitations for both linear and nonlinear systems. In order to examine the efficacy of the proposed algorithm as a suitable candidate for baseline-free SHM, numerical simulations are conducted for damage detection to the aforementioned 5 DOF B-W system for varying levels of nonlinearity change. Additionally, a numerically simulated 7 DOF B-W system is adopted in order to assess the performance of the proposed scheme against the increase in the DOF. This provides a test bed to evaluate the robustness of the methodology for a more practical case in which tall, flexible structures might be encountered and the detection results need to be established with the ground truth. The detailed theoretical derivations are presented in the ensuing sections.

6.2 Background

CCA is an exploratory method for determining the relationship between two multivariate sets of vectors by extracting information from the *cross-covariance* matrices. Developed in the late 1930's by Hoteling, CCA is a standard tool in statistical analysis that finds its application in econometrics, meteorology, medical studies and in the fields of signal processing as well. To the best of the knowledge of the author, the present work is one of the cardinal attempts at integrating CCA into a structural damage detection framework. The method finds two bases, one for each variable, that are optimal with respect to correlations and simultaneously obtains the corresponding correlations. Mathematically, it finds two basis vectors in which the correlation matrix between the variables is diagonal and mutually maximized by projecting the original set of variables onto an optimum subspace. An important feature of the method is that the dimensionality of these new bases is equal to or less than the dimensionality of the individual variables. In order to utilize the concepts of CCA into a structural dynamics framework, consider two column vectors $X = (x_1, x_2, ..., x_n)^T$ and $Y = (y_1, y_2, ..., y_n)^T$ of random variables with finite moments. Projecting these variables onto the basis vectors, the linear combinations $\mathcal{X} = X^T \hat{w}_x$ and $\mathcal{Y} = Y^T \hat{w}_y$ need to be mutually maximized which is given by:

$$\begin{aligned} \rho &= \frac{E[\mathcal{X}\mathcal{Y}]}{\sqrt{E[\mathcal{X}^2]E[\mathcal{Y}^2]}} \\ &= \frac{E[\hat{w}_x^T X Y^T \hat{w}_y^T]}{\sqrt{E\left[\hat{w}_x^T X X^T \hat{w}_x\right] E\left[\hat{w}_y^T Y Y^T \hat{w}_y\right]}} \\ &= \frac{\mathbf{w}_x^T \mathbf{C}_{xy} \mathbf{w}_y}{\sqrt{\mathbf{w}_x^T \mathbf{C}_{xx} \mathbf{w}_x \mathbf{w}_y^T \mathbf{C}_{yy} \mathbf{w}_y]}} \end{aligned} \tag{6.1}$$

It is a well understood concept that the covariance matrix of the random variables (centered to zero mean) can be expressed as:

$$\mathbf{C} = \begin{bmatrix} \mathbf{C}_{xx} & \mathbf{C}_{xy} \\ \mathbf{C}_{yx} & \mathbf{C}_{yy} \end{bmatrix} \tag{6.2}$$

The covariance matrix is essentially a block matrix where $\mathbf{C}_{xx}$ and $\mathbf{C}_{yy}$ are the within-set covariance matrices of $\mathcal{X}$ and $\mathcal{Y}$ respectively and the relation $\mathbf{C}_{xy} = \mathbf{C}_{yx}^T$ holds true for the cross-set covariance matrices. The aim of this methodology is to obtain the canonical correlations between $\mathcal{X}$ and $\mathcal{Y}$, that can be easily obtained by solving the following sets of eigen decomposition equations:

$$\left.\begin{array}{l} \mathbf{C}_{xx}^{-1}\mathbf{C}_{xy}\mathbf{C}_{yy}^{-1}\mathbf{C}_{yx}\hat{\mathbf{w}}_x = \rho^2\hat{\mathbf{w}}_x \\ \mathbf{C}_{yy}^{-1}\mathbf{C}_{yx}\mathbf{C}_{xx}^{-1}\mathbf{C}_{xy}\hat{\mathbf{w}}_y = \rho^2\hat{\mathbf{w}}_y \end{array}\right\} \tag{6.3}$$

where the eigenvalues ρ^2 are the squared *canonical correlations* and the corresponding eigenvectors $\hat{\mathbf{w}}_x$ and $\hat{\mathbf{w}}_y$ represent the normalized canonical correlation *basis vectors*. In practice, Eq. 6.3 can be recast into a single eigenvalue equation, given by:

$$\mathbf{B}^{-1}\mathbf{A}\hat{\mathbf{w}} = \rho\hat{\mathbf{w}} \tag{6.4}$$

where the matrices involved in the eigenvalue decomposition (EVD) are obtained as:

$$\mathbf{A} = \begin{bmatrix} \mathbf{0} & \mathbf{C}_{xy} \\ \mathbf{C}_{yx} & \mathbf{0} \end{bmatrix}, \quad B = \begin{bmatrix} \mathbf{C}_{xx} & \mathbf{0} \\ \mathbf{0} & \mathbf{C}_{yy} \end{bmatrix} \quad and \quad \hat{\mathbf{w}} = \begin{pmatrix} \mu_x\hat{\mathbf{w}}_x \\ \mu_y\hat{\mathbf{w}}_y \end{pmatrix} \tag{6.5}$$

It is worthy noting that a slightly different structure of the covariance matrix will give rise to the *traditional PCA* based approach, a concept that has consistently exhibited its potential in identifying the spatial and temporal patterns of structural damage, over the recent years. PCA, a special instance of CCA, can be formulated on substituting the matrices **A** and **B** as: $\mathbf{A} = \mathbf{C}_{xx}$ *and* $\mathbf{B} = \mathbf{I}$. PCA is a dimensionality reduction approach that projects the data onto the principal subspace, such that the variance of the projected data is maximized. In carrying out the projection, PCA reveals some simplified structures relevant to the dataset could be obtained by EVD on the covariance matrix. Prior to getting into the details of the proposed methodology, it is imperative to review the concepts of PCA through a structural dynamics viewpoint, profoundly discussed in *Chapter 2*. To tailor the basic CCA into a recursive framework, it becomes essential to revisit certain key theoretical developments that center around covariance estimates and FOEP techniques. In the backdrop of recursive implementation, the detailed formulation of FOEP presented in *Chapter 2*, provides an in-depth know how about incorporating eigenspace update through the rank-one perturbation theory. The readers are advised to acquaint themselves with the FOEP formulations previously described, prior to getting into the theoretical attributes of the present recursive approach.

6.3 RCCA: Detailed derivation

The use of CCA as a structural damage detection tool is a relatively nascent topic that has garnered certain attention among the researchers. The major drawback of using CCA is that it analyzes data in batches, primarily rendering it an offline approach. In order to detect finer levels of damage, it is essential to estimate the eigenspace recursively, thereby providing iteartive updates at each instant of time. The formulation of the basic CCA approach necessitates that the covariance matrix obtained from the set of recorded responses is essentially a block matrix, that needs to be updated for its online implementation. The major challenge towards tailoring the basic CCA towards an online implementation is to perform EVD on the block covariance matrix, which is particularly cumbersome and memory consuming. This can be alleviated through the use of the FOEP based approach which provides recursive updates of the eigenspace obtained from the previous eigenspace of the data at a particular instant. An important aspect to be considered here is that the FOEP based recursive approach updates the eigenspace characteristics at each time instant, instead of updating the covariance matrix as a whole, thereby reducing the time complexity of the recursive implementation.

The theoretical development of RCCA is premised on the objective of finding a recursive update of the eigenspace characteristics at each instant of time, a central idea envisaged from FOEP improvisation. The individual lower dimensional matrices shown in Eq. 6.2 need to be updated at each time stamp, providing recursive eigenspace estimates through the FOEP strategy. In this context, the response covariance matrix $\mathbf{C}_{xx}(k)$ at any instant k can be expressed as a function of the covariance matrix at the previous time stamp, $\mathbf{C}_{xx}(k-1)$ and the response vector X_k at the kth instant, according to:

$$\mathbf{C}_{xx}(k) = \frac{k-1}{k}\mathbf{C}_{xx}(k-1) + \frac{1}{k}X_k X_k^T \tag{6.6}$$

As previously mentioned, the FOEP based approach is a versatile idealization that does not take into consideration the nature of the dataset. The applicability of the method for nonstationary datasets involves the consideration of the recursive mean at each instant of time. The recursive mean update at the kth instant, μ_k, depends on the mean at the previous time stamp through the relation: $\mu_k = \frac{k-1}{k}\mu_{k-1} + \frac{1}{k}X_k$. The covariance estimate for cases involving mean updates can be expressed as:

$$\tilde{\mathbf{C}}_{xx}(k) = \frac{k-1}{k}\boldsymbol{\Sigma}_k^{-1}\boldsymbol{\Sigma}_{k-1}\tilde{\mathbf{C}}_{xx}(k-1)\boldsymbol{\Sigma}_{k-1}\boldsymbol{\Sigma}_k^{-1} + \boldsymbol{\Sigma}_k^{-1}\Delta\mu_k\Delta\mu_k^T\boldsymbol{\Sigma}_k^{-1} + \frac{1}{k}\left[\mathbf{X}_k - \mu_k\right]\left[\mathbf{X}_k - \mu_k\right]^T \tag{6.7}$$

The eigen decomposition of the covariance estimate shown in Eq. 6.7 can be written in terms of $\tilde{\mathbf{C}}_{xx}(k) = \tilde{\mathbf{W}}_k\tilde{\Upsilon}_k\tilde{\mathbf{W}}_k^T$. Equation 6.7 can be recast as:

$$\tilde{\mathbf{W}}_k\tilde{\Upsilon}_k\tilde{\mathbf{W}}_k^T = \frac{k-1}{k}\boldsymbol{\Sigma}_k^{-1}\boldsymbol{\Sigma}_{k-1}\tilde{\mathbf{C}}_{xx}(k-1)\boldsymbol{\Sigma}_{k-1}\boldsymbol{\Sigma}_k^{-1} + \boldsymbol{\Sigma}_k^{-1}\Delta\mu_k\Delta\mu_k^T\boldsymbol{\Sigma}_k^{-1} + \frac{1}{k}\left[\mathbf{X}_k - \mu_k\right]\left[\mathbf{X}_k - \mu_k\right]^T \tag{6.8}$$

The POC vector at the kth time instant can be estimated as: $\tilde{\psi}_k = \tilde{\mathbf{W}}_{k-1}^T X_k$. Carrying out proper substitutions in Eq. 6.8, one obtains:

$$\tilde{\mathbf{W}}_k \tilde{\Upsilon}_k \tilde{\mathbf{W}}_k^T = \frac{k-1}{k} \mathbf{\Sigma}_k^{-1} \mathbf{\Sigma}_{k-1} \tilde{\mathbf{W}}_{k-1} \tilde{\Upsilon}_{k-1} \tilde{\mathbf{W}}_{k-1}^T \mathbf{\Sigma}_{k-1} \mathbf{\Sigma}_k^{-1} + \mathbf{\Sigma}_k^{-1} \Delta\mu_k \Delta\mu_k^T \mathbf{\Sigma}_k^{-1} \quad (6.9)$$

$$+\frac{1}{k}\left[\mathbf{W}_{k-1}\tilde{\psi}_k - \mu_k\right]\left[\mathbf{W}_{k-1}\tilde{\psi}_k - \mu_k\right]^T \quad (6.10)$$

Scaling the data to unit variance, the above equation translates to:

$$\tilde{\mathbf{W}}_k \tilde{\Upsilon}_k \tilde{\mathbf{W}}_k^T = \frac{k-1}{k} \tilde{\mathbf{W}}_{k-1} \tilde{\Upsilon}_{k-1} \tilde{\mathbf{W}}_{k-1}^T + \Delta\mu_k \Delta\mu_k^T + \frac{1}{k}\left[\mathbf{W}_{k-1}\tilde{\psi}_k - \mu_k\right]\left[\mathbf{W}_{k-1}\tilde{\psi}_k - \mu_k\right]^T \quad (6.11)$$

For structural systems in particular, data is assumed to consistently evolve from zero mean processes. Therefore, the above equation can be simplified as:

$$\tilde{\mathbf{W}}_k k \tilde{\Upsilon}_k \tilde{\mathbf{W}}_k^T = \tilde{\mathbf{W}}_{k-1}\left[(k-1)\tilde{\Upsilon}_{k-1} + \tilde{\psi}_k \tilde{\psi}_k^T\right] \tilde{\mathbf{W}}_{k-1}^T \quad (6.12)$$

An important conclusion that could be drawn from the aforementioned discussion is the fact that the term $[(k-1)\tilde{\Upsilon}_{k-1} + \tilde{\psi}_k \tilde{\psi}_k^T]$ should be diagonally dominant, for the RPCA algorithm to be stable and robust. Subsequently, as the system consists of very low to moderate levels of damping, the EVD can be evaluated using the Gershgorin's theorem [24]. The term $\tilde{\psi}_k \tilde{\psi}_k^T$ represents the correlation between the POC estimates at a particular instant. Substituting from Eq. 5.3, the covariance between POC estimates can be written as (to an arbitrary scale factor):

$$\begin{aligned} \tilde{\psi}_k \tilde{\psi}_k^T &= \tilde{\mathbf{W}}_{k-1}^T X_k {X_k}^T \tilde{\mathbf{W}}_{k-1} \\ &= q_{k-1} {q_{k-1}}^T + \gamma {q_{k-1}}^T + \gamma^T q_{k-1} + \gamma\gamma^T \end{aligned} \quad (6.13)$$

As far as the dynamics of structural systems are considered, the error term in the Eq. 6.13 can be neglected as the number of sampling points increases and under moderate to low damping [24]. Recognize the similarity of the first term in Eq. 6.13 that closely resembles the covariance of the normal coordinates at the instant $(k-1)$. Thus the term $q_{k-1} {q_{k-1}}^T$ represents a matrix whose diagonal terms dominate its off-diagonal terms; hence, the term $\tilde{\psi}_k \tilde{\psi}_k^T$ can be safely assumed to be diagonally dominant. This in turn, ensures the diagonal dominance of the matrix $\left[(k-1)\tilde{\Upsilon}_{k-1} + \tilde{\psi}_k \tilde{\psi}_k^T\right]$, facilitating straightforward application of Gershgorin's theorem. Hence for a structural system, the recursive eigenspace update is obtained using the FOEP approach which provides a less computationally intensive algorithm in a recursive framework. The EVD of the matrix $\left[(k-1)\tilde{\Upsilon}_{k-1} + \tilde{\psi}_k \tilde{\psi}_k^T\right]$ can be substituted as $\mathbf{H}_k \mathbf{\Lambda}_k \mathbf{H}_k^T$ into Eq. 5.8 as,

$$[\tilde{\mathbf{W}}_k][k\tilde{\Upsilon}_k][\tilde{\mathbf{W}}_k]^T = [\tilde{\mathbf{W}}_{k-1}\mathbf{H}_k][\mathbf{\Lambda}_k][\tilde{\mathbf{W}}_{k-1}\mathbf{H}_k]^T \quad (6.14)$$

yielding the following iterative update equations:

$$\left.\begin{aligned}\tilde{\mathbf{W}}_k &= \tilde{\mathbf{W}}_{k-1}\mathbf{H}_k \\ \boldsymbol{\Upsilon}_k &= \tfrac{\boldsymbol{\Lambda}_k}{k}\end{aligned}\right\} \tag{6.15}$$

The recursive algorithm of Eq. 6.6 is transformed to obtain the values of H_k and Λ_k. As previously described, the term $(k-1)\tilde{\boldsymbol{\Upsilon}}_{k-1} + \tilde{\psi}_k\tilde{\psi}_k^T$ is diagonally dominant, which ensures that the eigen values are the diagonal entries of the matrix. therefore, the ith diagonal entry of the term Λ_k can be represented as $(k-1)\lambda_i + \tilde{\psi}_i^2$, where λ_i is the (i,i) element of $\tilde{\boldsymbol{\Upsilon}}_{k-1}$ and $\tilde{\psi}_i$ is the ith entry of the POC estimate. Once the eigen values are obtained, the corresponding eigen vectors can be found out, leading to H_k.

Based on the above discussion, the recursive eigenspace estimate of a single set of responses (more particularly, the responses obtained in a single direction) is achieved. The block covariance matrix shown in Eq. 6.2 consists of the covariance matrices obtained from the responses in the orthogonal direction as well. Further, the block matrix comprises of the cross covariance matrices, that need to be recursively updated as well. In line with the above derivations, a similar presentation of the recursive eigen estimation can be arrived at, which is repetitive and therefore, omitted here for brevity. Once each of the individual covariance estimates are obtained, the next task is to implement these matrices onto a recursive framework prescribed by the eigenvalue problem shown in Eq. 6.4.

At any time instant k, the fundamental matrices involved in the eigen decomposition of Eq. 6.4 can be expressed as:

$$\mathbf{A}(k) = \begin{bmatrix} \mathbf{0} & \mathbf{C}_{xy}(k) \\ \mathbf{C}_{yx}(k) & \mathbf{0} \end{bmatrix}, \quad \mathbf{B}(k) = \begin{bmatrix} \mathbf{C}_{xx}(k) & \mathbf{0} \\ \mathbf{0} & \mathbf{C}_{yy}(k) \end{bmatrix} \tag{6.16}$$

The recursive updates of the covariance estimates can be substituted in to the above equation. These updates follow a structure similar to the one shown in Eq. 6.6 and can be expressed as:

$$\left.\begin{aligned}\mathbf{A}(k) &= \begin{bmatrix} \mathbf{0} & \frac{k-1}{k}\mathbf{C}_{xy}(k-1) + \frac{1}{k}X_kY_k^T \\ \frac{k-1}{k}\mathbf{C}_{yx}(k-1) + \frac{1}{k}Y_kX_k^T & \mathbf{0} \end{bmatrix} \\ \mathbf{B}(k) &= \begin{bmatrix} \frac{k-1}{k}\mathbf{C}_{xx}(k-1) + \frac{1}{k}X_kX_k^T & \mathbf{0} \\ \mathbf{0} & \frac{k-1}{k}\mathbf{C}_{yy}(k-1) + \frac{1}{k}Y_kY_k^T \end{bmatrix}\end{aligned}\right\} \tag{6.17}$$

An important feature of the FOEP based techniques is their ability to reduce time complexities by a significant margin. As evident from the literature, the step wise update of the covariance matrix at each iteration is cumbersome [25] and calls for efficient techniques to perform the recursive updates. Additionally, performing an iterative estimation in the backdrop of a generalized eigenvalue problem is more involved and complicated. A simplified route that could be envisioned is through the structural modification of the equation under study and then performing the recursive updates. Based on this premise, consider a simplified version of Eq. 6.4 at the kth time instant, given by:

$$\mathbf{A}(k)\hat{\mathbf{w}}(k) = \rho(k)\mathbf{B}(k)\hat{\mathbf{w}}(k) \tag{6.18}$$

From a thorough understanding of the FOEP approach, it is clear that Eq. 6.18 needs to be updated for every sample. As evident from Eq. 6.15, the rank-one update of the covariance matrix provides the recursive estimations at each instant. Similar to this notion, consider the rank-one perturbation of the eigen decomposition given by Eq. 6.18:

$$(\mathbf{A}+\Delta\mathbf{A})(\hat{\mathbf{w}}+\Delta\hat{\mathbf{w}}) = (\rho+\Delta\rho)(\mathbf{B}+\Delta\mathbf{B})(\hat{\mathbf{w}}+\Delta\hat{\mathbf{w}}) \tag{6.19}$$

Considering the fact that the perturbed matrices are of low magnitude, each of the small perturbations correspond to evaluating the matrix at the $k+1^{th}$ instant. This translates to evaluating the EVD at the $k+1^{th}$ time stamp, which is precisely the basic objective of the FOEP methodology:

$$\mathbf{A}(k+1)\hat{\mathbf{w}}(k+1) = \rho(k+1)\mathbf{B}(k+1)\hat{\mathbf{w}}(k+1) \tag{6.20}$$

From the aforementioned discussions, it is obvious that the individual matrices involved in the EVD can be recursively updated following the similar notion expressed in Eq. 6.17. A key feature of this update is the recursive estimation of the eiegnspace, which can be readily obtained by following the same methodology as in Eq. 5.11. This demonstrates that the eigenspace updates can be obtained through the FOEP technique without actually performing EVD on the subsequent data points, thereby reducing the time complexity and memory consumption, to an enormous extent.

$$\begin{aligned}
&\begin{bmatrix} \mathbf{0} & \frac{k}{k+1}\mathbf{C}_{xy}(k)+\frac{1}{k+1}X_{k+1}Y_{k+1}^T \\ \frac{k}{k+1}\mathbf{C}_{yx}(k)+\frac{1}{k+1}Y_{k+1}X_{k+1}^T & \mathbf{0} \end{bmatrix}(k+1)\left[\mathbf{W}_k\mathbf{H}_{k+1}\right] \\
&= \Lambda_{k+1}\begin{bmatrix} \frac{k}{k+1}\mathbf{C}_{xx}(k)+\frac{1}{k+1}X_{k+1}X_{k+1}^T & \mathbf{0} \\ \mathbf{0} & \frac{k}{k+1}\mathbf{C}_{yy}(k)+\frac{1}{k+1}Y_{k+1}Y_{k+1}^T \end{bmatrix}\left[\mathbf{W}_k\mathbf{H}_{k+1}\right] \\
&= \frac{\Lambda_{k+1}}{k+1}\left\{\begin{bmatrix} k\mathbf{C}_{xx}(k) & \mathbf{0} \\ \mathbf{0} & k\mathbf{C}_{yy}(k) \end{bmatrix}+\begin{bmatrix} X_{k+1}X_{k+1}^T & \mathbf{0} \\ \mathbf{0} & Y_{k+1}Y_{k+1}^T \end{bmatrix}\right\}\left[\mathbf{W}_k\mathbf{H}_{k+1}\right]
\end{aligned} \tag{6.21}$$

6.4 Recursive damage sensitive features

The current framework exploring the concepts of RCCA facilitates online processing of the data and yields the recursive updates of the eigenvalues and the eigenvectors, referred to as the 'eigenspace', in a pedantic sense. The eigenspace by itself is inadequate in exhibiting the deviations inflicted at the onset of the damage. Therefore, certain damage markers, known as DSFs, are employed to ascertain the presence of damage and its location, visually or otherwise. The key characteristics of a good DSF lies in its potential to detect the presence of damage, locate and estimate the severity of damage and effectively distinguish between the damaged and undamaged states of the structure. Further, the auxiliary attributes of these DSFs arise from their ability to function online, in order to identify the change of states in real time. In this context, the aforementioned TVAR coefficients and the RRE vectors are utilized as appropriate DSFs for identifying the spatio-temporal patterns of damage in time. The detailed derivation for the same has already been encountered in *Chapter 3* and hence, not reported here for brevity.

6.5 Proposed algorithm

The comprehensive methodology followed in this chapter for identifying and localizing the damage in real time entails two contributing modules: *temporal* and the *spatial* module, working simultaneously in a single framework. The primary course of action followed is to first identify the instant of damage, and then proceed on to detecting its location. The first module deals with identifying the instant of damage where the recorded vibration data is provided as input to the RCCA algorithm, in order to obtain the transformed responses. These responses, consequently modeled using TVAR, produce TVAR coefficients which are indicative of damage to the system. On ascertaining the instant of damage, the spatial module is invoked where the spatial RRE is tracked over each DOF to identify the location of damage. A distortion in the RRE plots confirms the exact location of damage in the structure. For an easy comprehension of the detailed process, the basic steps of the algorithm are enumerated as follows:

1. First, a block covariance matrix incorporating the individual auto- and cross-covariance matrices is fabricated from the set of physical responses. Traditional CCA is applied on some initial data points (around 100 in number) in order to estimate the initial eigenspace. The number of data points chosen here is arbitrary and considered only to stabilize the algorithm for subsequent real time damage detection. The batch implementation comprises the EVD updates shown in Eq. 6.3, yielding the eigenspace represented by ρ and $\hat{\mathbf{w}}$.

2. The recursive gain depth parameter is employed to estimate the covariance estimate at the present time instant using the covariance estimate at the preceding time instant. From these recursive updates, the eigenvector and eigenvalue matrices are updated using the FOEP approach and the trans-

formed responses (principal components) are obtained using the RCCA algorithm. While the covariance updates for the matrices are given by Eq. 6.17, the iterative set of eigenspace updates carried out after the initial batch implementation can be easily obtained through Eq. 5.11.

3. The transformed responses are fit using TVAR models of appropriate model order. This provides a set of TVAR coefficients that evolve with the progression of time and aid in identifying the exact instant of damage, governed by Eq. 5.16. In addition, the temporal RRE plot, given by Eq. 4.16, further substantiates the damage instant corroborated by the TVAR plots.

4. Once the instant of damage is determined, the algorithm shifts on to the next module where the spatial detection of damage takes place. The mathematical expressions administering its implementation can be clearly obtained from Eq. 4.19. The local RRE is tracked over the entire system as a whole that generates distortions in the plot, indicative of damage at that particular location.

The aforementioned steps for the proposed method are shown in the form of a stepwise flowchart in Fig. 6.1. Based on the theoretical derivations, the

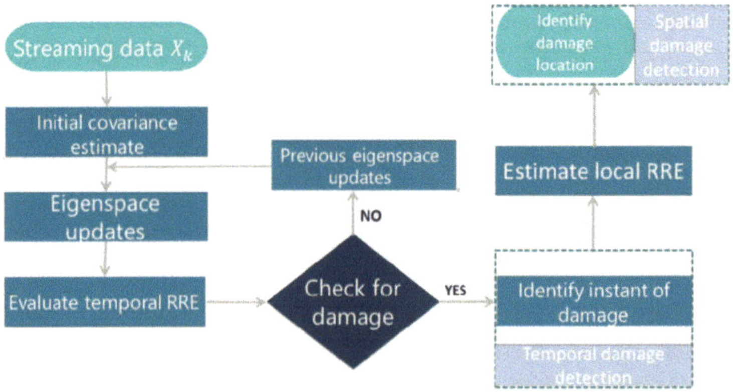

Figure 6.1: Flowchart of the RCCA method.

TVAR coefficients and RRE show deviations in the mean level of the plots, that visually aid in identifying the instant of damage. A few important characteristics of the proposed algorithm include: (i) implementation on the data obtained at each time stamp, as and when the data becomes available; signifying that the algorithm essentially operates *online* (ii) detection of damage instant and its location without ad-hoc windowing or gathering of any sort of baseline data, making it *baseline-free* and (iii) absence of parameters controlling the functioning of the algorithm, rendering it *parameter independent*. These characteristics of the proposed algorithm makes it an ideal candidate for real time structural damage detection framework.

6.6 Detection results using proposed algorithm

Case studies are undertaken for both global and local damage detection using the proposed algorithm in the context of numerical simulations and verified using experimental studies. In this context, two numerically simulated systems are considered here: (i) a 5 DOF B-W system and (ii) a 7 DOF B-W system. The source of nonlinearity for both the systems is through the nonlinear force parameter at the base of the models, controlled by the nonlinear force term κ. It can be conjectured at this stage that the effect of change of the value of κ for the latter case will not be as pronounced as compared to the 5 DOF model, primarily due to the increasing DOF of the structure, that constraints the nonlinearity propagation throughout the model. The temporal damage detection cases are obtained by changing the value of the nonlinear force term κ for the B-W system at a specified instant of time, sequentially followed by the cases for real time spatial damage detection.

6.6.1 Temporal damage detection studies for the B-W systems excited using white noise

A brief numerical study on aforementioned systems has been carried out using Gaussian white noise as an excitation. Temporal damage detection cases are studied first by sequentially changing the κ values corresponding to 15% and 10% changes in nonlinear characteristics respectively.

6.6.1.1 Temporal damage detection results for the 5 DOF B-W system

Case studies using 5 DOF B-W system for a 10% and 15% change in nonlinearity are reported in this section. Recent investigation has revealed that damages of the order of 15% have often been reported as a lower bound for real time vibration based damage detection. However, the proposed algorithm provides successful

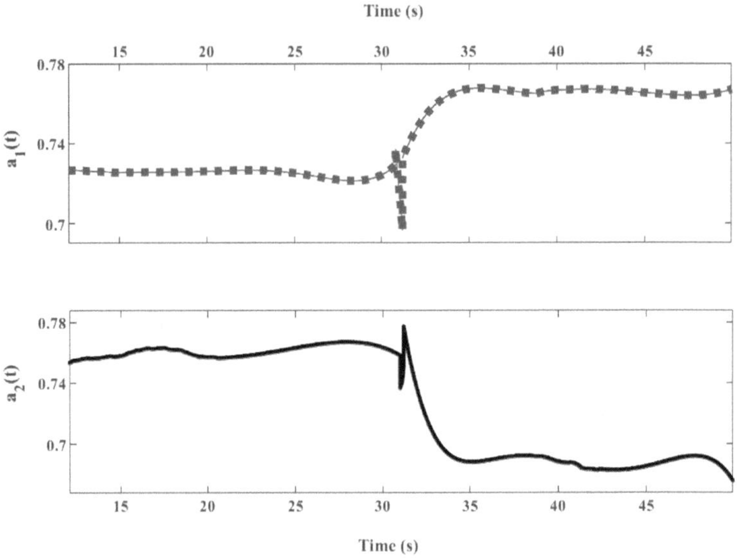

Figure 6.2: Recursive DSF plot for 15% temporal damage for the 5 DOF B-W system (using white noise excitation).

detection results for finer levels of damage, as low as 10%, for the 5 DOF B-W system. The detection results using the proposed DSFs are presented next in detail.

From Fig. 6.2 a damage instant at 31s can be clearly observed for 15% change in nonlinearity. The sharp peak discerned at the damage instant primarily attributes to the sudden change in the mean level of the TVAR plot, an event that is consistent for both the coefficients. In order to validate these findings, the temporal RRE plot clearly substantiates the instant of damage through a sharp distortion at 31s. A separate figure dedicated towards real time damage detection for a 10% change in nonlinearity is shown in Fig. 6.3. From both Figs. 6.2 and 6.3 it can be observed that, the plots of the TVAR coefficients show changes in the mean level of the plot around the 31s mark, indicating a possible damage event. Further, the residual error when examined in a recursive framework, detects the instant of damage at 31s, clearly observed from the figures, thereby validating the inferences concluded from the TVAR plots. In line with the above findings, it can be very well established that the proposed RCCA algorithm is efficient in determining fine levels of damage in real time, a feat that has only been reported in a recently established hybrid algorithm using an association of the RPCA-RSSA strategies.

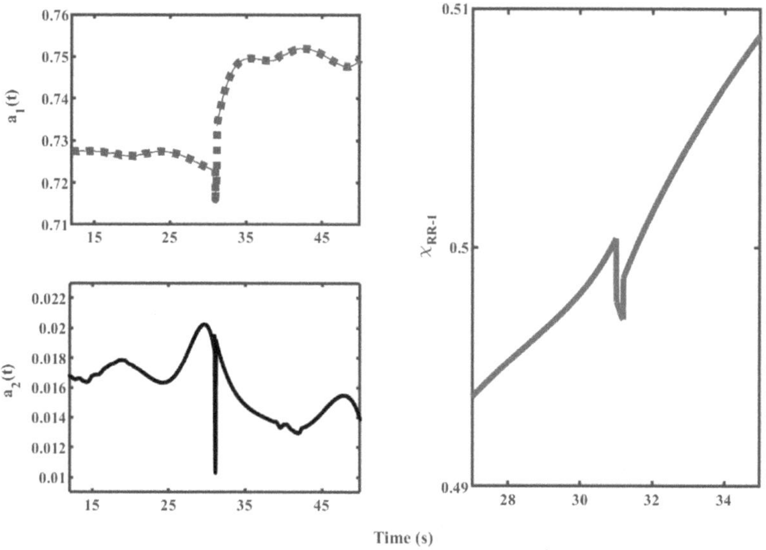

Figure 6.3: Recursive DSF plot for 10% temporal damage for the B-W system (using white noise excitation).

6.6.1.2 Spatial damage detection results for the Bouc-Wen 5 dof model

The present section deals with the performance of the proposed algorithm towards simultaneously detecting a spatio-temporal damage in real time, which is one of the key entitlements of the current work. The notion of local damage is brought about by a change in linear storey stiffness of a particular DOF for a MDOF system, that forms the key cornerstone for processing online spatio-temporal damage. It is worth noting that the proposed algorithm invokes the spatial module only when the instant of damage has been previously ascertained in the temporal constituent. A through investigation of the recently established real time damage detection literature reveals that detection studies using the RPCA algorithm could be traced back to a lower limit of 25% in real time. Motivated by this key shortcoming, the proposed method focuses on the detection potential for the weakly nonlinear B-W system for a spatial damage of both 10% and 15%, the results of which are presented next in detail. The key consideration to be incorporated here is the fact that the system is made fully nonlinear by scaling up the value of the nonlinear parameter κ to unity and the linear storey stiffness of the third floor is reduced by 10% and 15%, respectively, at 31s from the start of the excitation.

From Fig. 6.4, it can be visualized that a clear damage instant is identified for 15% reduction in storey stiffness through the deviation in the TVAR plot. After the damage instant is established from the TVAR plot, the spatial RRE in the

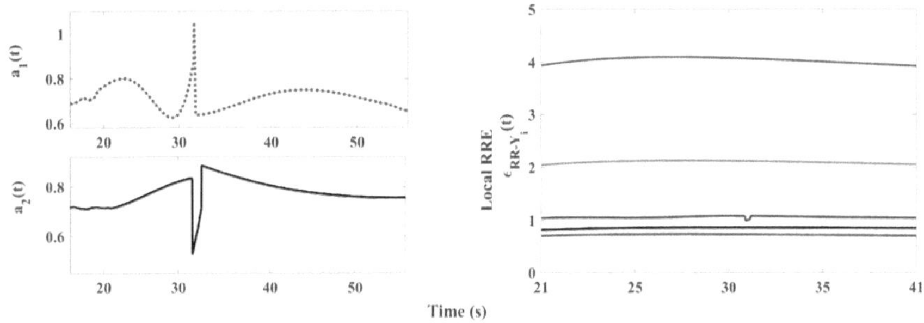

Figure 6.4: Recursive DSF plot for 15% spatio-temporal damage for the B-W system.

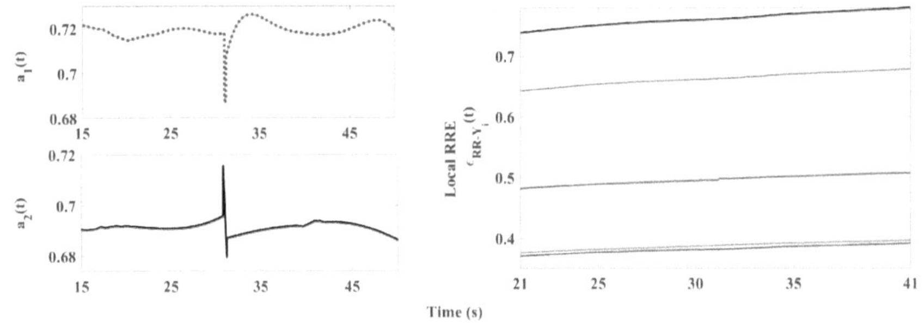

Figure 6.5: Recursive DSFs for 10% spatio-temporal damage for the B-W system (under white noise excitation).

vicinity of damage (say 21s-41s) is examined. The spatial RRE plots for all the DOFs are portrayed beside the TVAR coefficients clearly indicating a distortion in the 3rd DOF, thereby conforming to the previous consideration of the occurrence of damage at 31s confined to that floor. Furthermore, a sharp change in the mean level of the TVAR plots at 31s is clearly evident from Fig. 6.5, thereby establishing the fact that the proposed method is well suited for identifying the local patterns of damage for a value as low as 10%, which is clearly a major accomplishment considering the online nature of the algorithm. Such a low percentage of spatio-temporal damage has only been reported in a recently established hybrid scheme incorporating RPCA-RSSA based methods. However, the key shortcoming of the said method is its time complexity as opposed to the proposed scheme that independently identifies such a finer aspect of damage. This is one of the major entitlements of the current study.

6.6.2 Detection results for the 5 DOF B-W system excited by El Centro ground motion

The robustness of the proposed RCCA algorithm towards solving a wide range of damage identification problems can ensured with its application towards a wide range of excitation that includes nonstationary cases as well. Some of the well published literature on damage detection have consistently failed to detect damage for nonstationary cases with a reasonable degree of confidence. This situation arises primarily due to the inherent formulation of the damage detection techniques which are mostly pivoted around linear systems excited using stationary events such as white noise. However, the proposed method has successfully identified real time damage for the aforesaid nonlinear system and is conjectured to provide good results for case studies involving nonstationary excitations as well. To this effect, the use of TVAR modeling on the transformed responses substantially aid in the damage detection scheme, by masking the nonstationary effect of the input excitation.

To ensure the versatility of the proposed RCCA algorithm, the B-W system is excited using El Centro ground motion and the detection results are presented. The damage to the model is numerically simulated through a 25% and a 20% change in the nonlinear force term at the base of the model, that eventually contributes to a global damage to the system. Recently established RPCA algorithm provides a real time detection of 30% for the El Centro excitation case, thereby confirming the superiority of the proposed RCCA based scheme over the previously established ones. In the subsequent sections, a thorough comparison of the two said methods will be provided, that sheds a light on the efficacy of the proposed algorithm over the recently established schemes. The detection results displayed in Figs. 6.6 and 6.7 clearly indicate the exact instant of damage at 25s from the start of the excitation. The temporal RRE plots, in addition to the TVAR coefficients, aptly validate the damage instant for both cases.

The local damage is simulated for the B-W system by numerically reducing the linear storey stiffness of the 3rd floor by 25% and 20%, respectively, at 25s from the commencement of the excitation. The plots of the TVAR coefficients shown in Fig. 6.8, indicate a clear instant of damage at 25s. The spatial module of the algorithm gets invoked at the determination of the damage instant, due to which, the spatial RREs are tracked at each of the floor levels. In the neighborhood of the damage instant, the spatial RREs are evaluated for each floor, displayed in Fig. 6.8, which show distortion at 25s for the 3rd storey. This confirms the localization of damage at the 3rd floor of the system, corroborated from the spatial RRE plots. Similar results are obtained from Fig. 6.9 that indicate spatio-temporal damage in real time for a 20% change in the linear storey stiffness. Recently established detection schemes such as RPCA have reported a lower limit

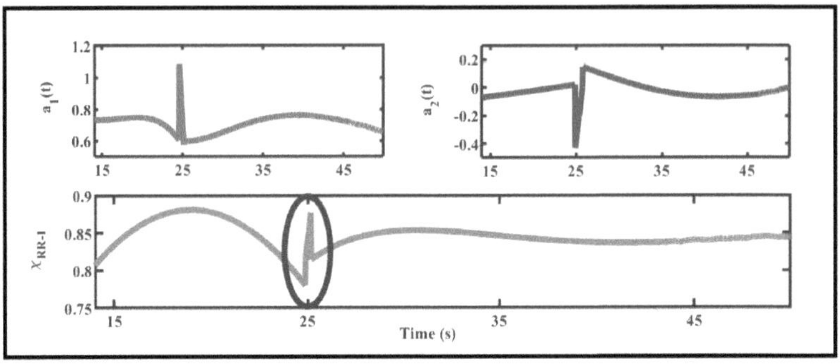

Figure 6.6: Recursive DSF plot for 25% temporal damage for the B-W system under El Centro excitation.

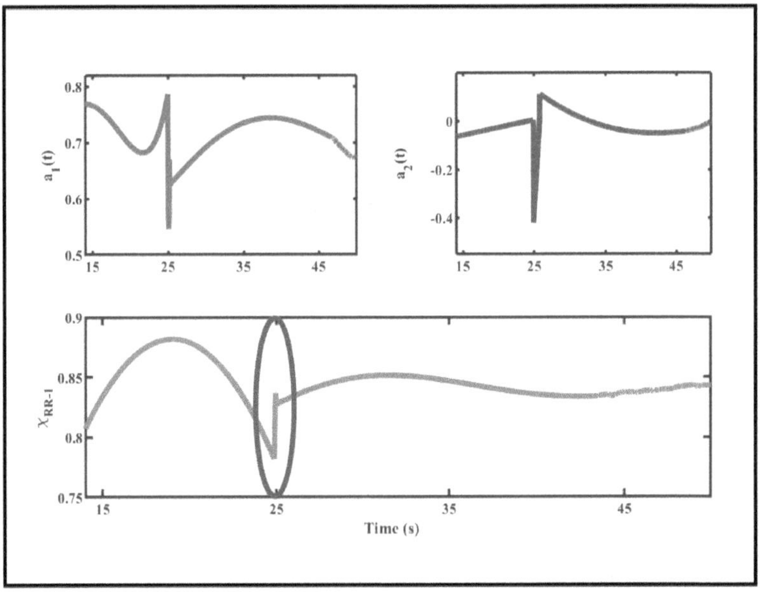

Figure 6.7: Recursive DSF plot for 20% temporal damage for the B-W system under El Centro excitation.

of detectability of *30% global damage* for a nonstationary input excitation; hence, a key contribution of the present work is the development of a framework that successfully reports *spatio-temporal damage* as low as *20%* for a nonstationary excitation.

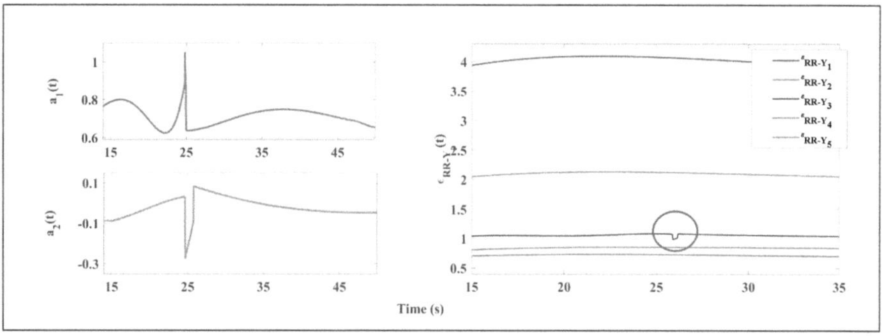

Figure 6.8: Recursive DSF plot for 25% spatio-temporal damage for the B-W system under El Centro excitation.

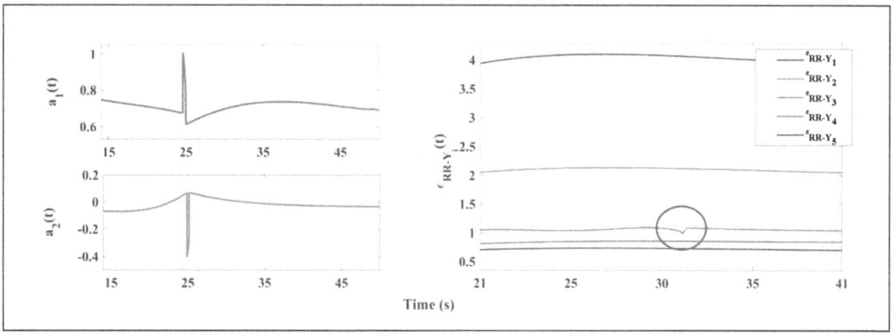

Figure 6.9: Recursive DSF plot for 20% spatio-temporal damage for the B-W system under El Centro excitation.

6.6.3 Case study for an under determined system

It is worth noting that the covariance matrix generated from the physical responses obtained from the structure are usually full rank in nature. In practical scenarios, cost, improper accessibility and unavailability of good quality sensors, impede the possible instrumentation of all the physical DOF of the structure. This results in an *under detrmined case*, where the number of instrumented sensors is less than the number of DOF, thereby fabricating a rank deficit matrix of the physical responses. The applicability of any damage detection scheme towards such continuous health monitoring and condition based maintenance lie in the performance assessment of the methods towards effectively identifying the damage patterns for under determined cases as well. In this context, the proposed method is examined in a recursive framework to effectively determine the instant and location of damage for such a case. The performance check mainly assesses the functioning of the algorithm to determine damage from a subset of sensors, considered below:

1. Case 1: Generating a block covariance matrix using a subset of sensors instrumented at the 3rd, 4th and 5th floors.

2. Case 2: Generating a block covariance matrix using the first set of responses obtained from the 3rd, 4th and 5th floors and the other set obtained from the 2nd, 3rd and 5th floors.

For both these case studies, the previously described B-W system is considered that undergoes a temporal damage of 20% and 15% at 31s from the start. From a theoretical standpoint, the number of instrumented sensors must equal the number of actively participating modes, which is 3 in the present context. Thus the resultant under determined system is expected to produce POMs and POVs corresponding to the reduced system order. Once the block covariance matrix is created, the RCCA algorithm operates on the reduced order physical responses to provide a set of corresponding transformed responses, on which the TVAR models are fit. In this context, two cases are considered: (i) a global damage considering 15% change in nonlinearity and (ii) a local damage case considering reduction of the linear storey stiffness by 20%, at 31s from the start of the excitation.

Detection results for Case 1:
The first case considers the streaming input data obtained from 3rd, 4th and 5th floors that is required for creating a block covariance matrix. Once the transformed responses are obtained, the TVAR modeling is adopted that provides the key DSFs for identification of damage. The TVAR plots shown in Fig. 6.10 depict the exact instant of damage by indicating a change in the mean level at 31s. In order to validate these findings, the temporal RRE plot shows a distortion that confirms the damage event occurring at 31s. It is obvious that the present results are invariant of any instabilities that might arise due to the under determined nature of the responses, thereby demonstrating the superior prowess of the proposed scheme over recently established methods such as RPCA, that are prone to instabilities, even for a 20% global change in nonlinearity.

Detection results for Case 2:
The present case provides an insight into the efficacy of the proposed scheme in determining spatio-temporal damage for an underdetermined case. The *first instance* considers the block covariance matrix to be formed by the set of physical responses obtained from 3rd, 4th and 5th floors. The linear stiffness of the 3rd storey is reduced by 20% at 31s from the start. The transformed responses are modeled using TVAR coefficients, portrayed in Fig. 6.11. The exact instant of damage is identified at 31s from TVAR plots. On determining the exact instant

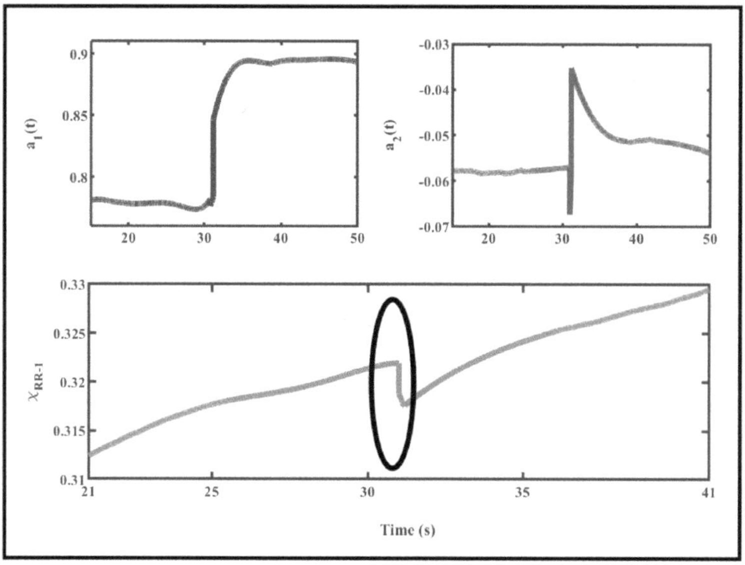

Figure 6.10: Recursive DSF plot for Underdetermined case-1, Global damage, 15% change in non-linearity excited by white noise.

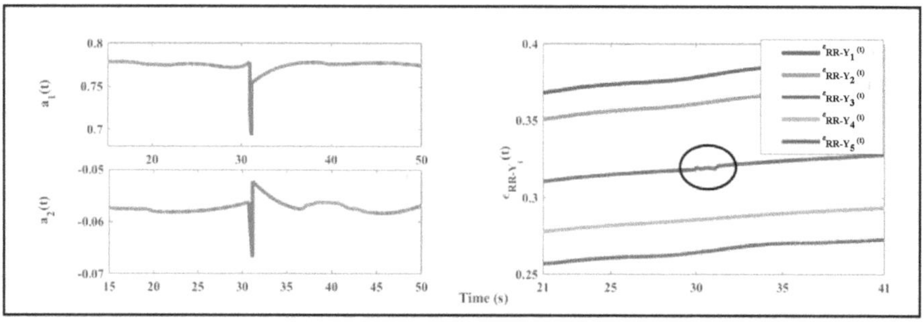

Figure 6.11: TVAR plot for Underdetermined case-2, first instance.

of damage, the spatial module of the algorithm is invoked where the local RRE for the reduced number of floors is tracked over time. It is clear from Fig. 6.11 that the local RRE corresponding to the 3rd floor provides a significant distortion at 31s. It can therefore be interpreted that the absence of a similar pattern of distortion from the other floor levels validates the exact localization of damage, a phenomenon indeed confined only to the 3rd storey of the system.

The *second instance* considers the inputs of two sets of responses: The first set obtained from the 3rd, 4th and 5th floor, while the other set is acquired from

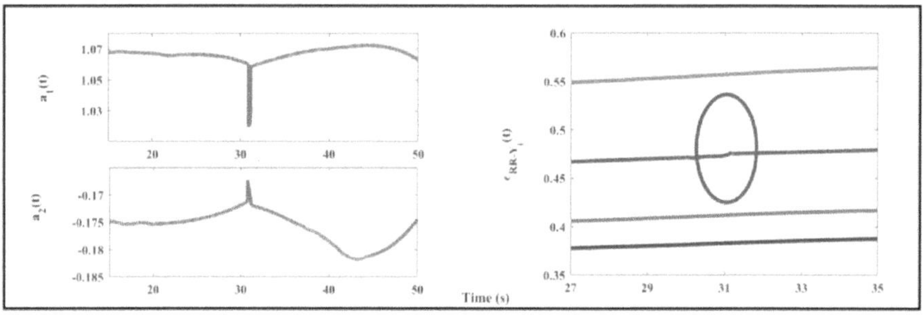

Figure 6.12: TVAR plot for Underdetermined case-2, second instance.

the 2nd, 3rd and 5th floors. The block covariance matrix is generated based on these inputs. It can be conjectured that the detectability of the proposed scheme under a non-identical set of responses might be affected, primarily due to the divergent nature of the input processes, which makes the individual covariance matrices dissimilar. The plots of recursive DSFs shown in Fig. 6.12 provide a clear insight into the matter. Although the exact damage instant of 31s can be observed from the TVAR plots, the spatial module fails to provide adequate detection results for localizing the damage in real time. It is clearly visible that the distortions in the 3rd DOF are not as noteworthy as the ones reported in the previous detection estimates, thereby rendering the spatial localization scheme inconsequential for this case. A review of the recently established RPCA based detection strategy provides a 20% and 30% lower limit of detectability for the global and local damage cases, respectively. Based on the above inferences, the proposed method successfully identifies comparatively finer levels of spatio-temporal damage, which is a key contribution, considering the real time essence of the algorithm.

6.6.4 *Case study of a 7 DOF B-W system excited using white noise*

A numerically simulated 7 DOF B-W system is excited using a Gaussian white noise of 50s duration, sampled at 100 Hz. An extension of the previously inspected 5 DOF B-W system, the key motivation for using this system for real time damage detection studies is to assess the performance of the method against an increase in the DOF. It is well understood that the nonlinear change at the base of the model propagates strongly to the vicinity of the damage. Therefore, the nonlinear propagation throughout the model is an arduous task and can effect only a few of the neighboring DOF of the superstructure. In order to clearly understand the aspects of damage for an increased number of DOF, keeping all other factors invariant, the proposed method is applied on to the model under consideration and

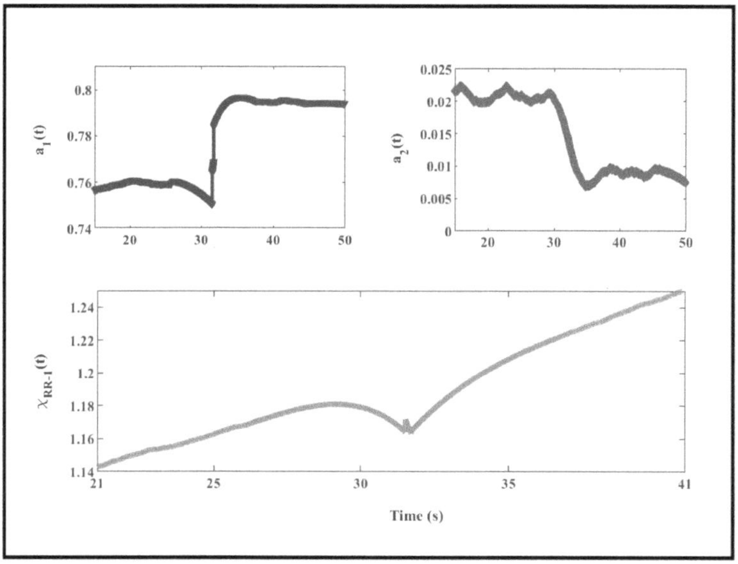

Figure 6.13: Recursive DSF plot for 20% damage for the B-W 7 DOF system under white noise excitation.

the detailed results are reported. In this context, the nonlinear force parameter κ is varied by 20% and 15%, individually, at 31s from the commencement of the excitation. The recursive DSFs shown in Fig. 6.13 clearly portray the exact instant of damage through a sharp change in the mean level at 31s. Additionally, Fig. 6.14 provides successful detection results for a 15% real time damage to the 7 DOF B-W system.

A salient feature of the DSFs can be observed here. A recently established real time damage detection algorithm, RPCA, has provided successful results for a 15% temporal damage case. The study, conducted on a 5 DOF B-W system failed to show significant distortions in the global RRE plots. The robustness of the DSFs lie in their ability to determine damage for a wide range of applications, including a varied class of nonlinear systems as well. One of the key features of the proposed scheme is to incorporate the same specific set of DSFs for a wide class of problems dealt within the scope of this study, without compromising on the time complexity or the efficacy of the proposed methodology. Therefore, the significant distortion in the RRE plots displayed in both the Figs. 6.13 and 6.14 is of key importance and provides extensive evidence towards the effectiveness of the proposed methodology and thereby, the DSFs.

The key entitlement of the work lies in the ability of the method to identify spatio-temporal damage in real time. Following a similar notion, the proposed

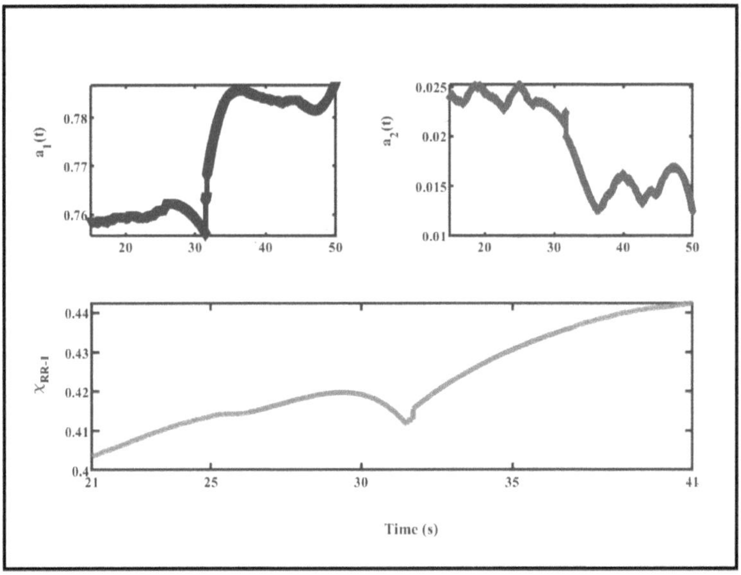

Figure 6.14: Recursive DSF plot for 15% damage for the B-W 7 DOF system under white noise excitation.

RCCA method is examined for a local damage case for the 7 DOF B-W system. From the previous detailed discussions, it is clear that the change in the linear storey stiffness remains confined to a single DOF; therefore, the proposed method is expected to provide credible results for spatio-temporal damage for the current model as well. Hence, it is envisioned that the RCCA scheme will provide successful detection results for a similar percentage of local damage in real time. In this regard, the linear storey stiffness of the 3rd floor is reduced by 15% and 10% and studied extensively in separate case studies. For the first case, the sharp peak of the TVAR plots shown in Fig. 6.15, indicates a clear damage instant at 31s. While the spatial RREs of all the other floors fail to indicate a distortion, the significant deviation that occurs in the 3rd storey of the system, validates the exact occurrence of damage, recursively, in a simultaneous framework. Detection results for a 10% linear storey stiffness change is provided in Fig. 6.16 that clearly brings out the spatio-temporal aspect of the proposed framework. Based on the above discussions, some of the key conclusions that can be drawn are:

1. The propagation of the change in nonlinearity for an increased number of DOF is comparatively less, a concept that is discussed in detail in the later stages of the chapter.

2. The use of RRE as an efficient DSF even for fine level of damage of the order of 10% is a key entitlement of the work, which readily provides

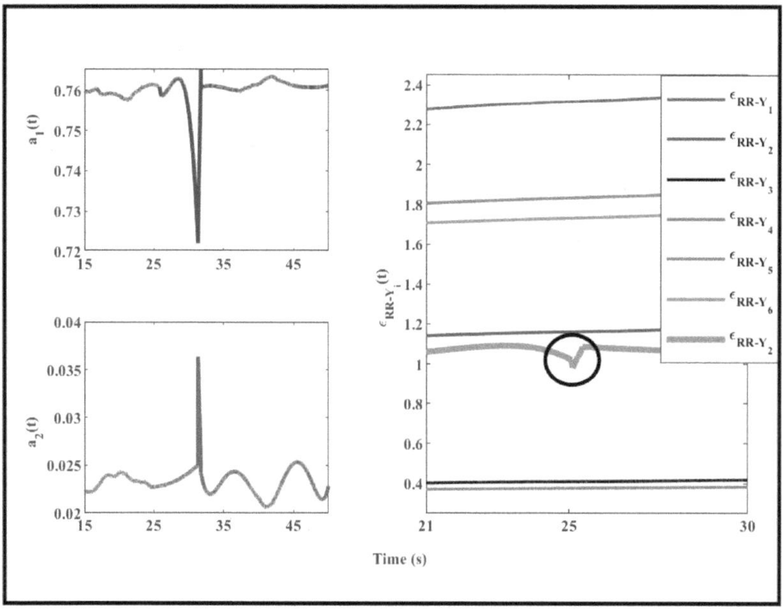

Figure 6.15: Recursive DSF plot for 15% spatio-temporal damage for the B-W 7 DOF system under white noise excitation.

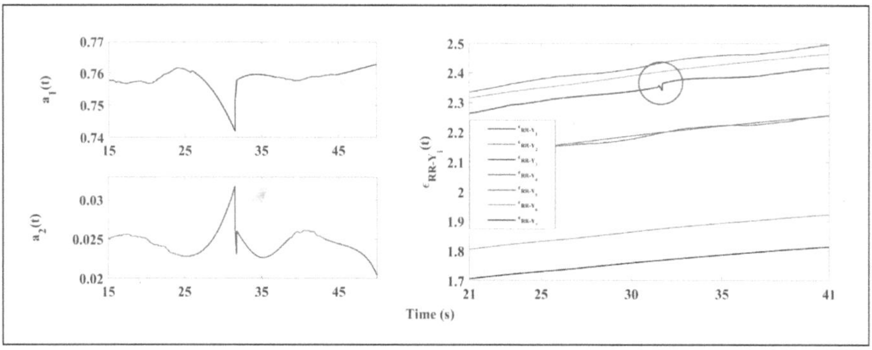

Figure 6.16: Recursive DSF plot for 10% spatio-temporal damage for the B-W 7 DOF system under white noise excitation.

an authentication to the efficacy of the proposed algorithm in addition to recognizing the utility of the said DSFs.

3. The algorithm provides almost similar results for local damage detection for both the above mentioned B-W systems. The purpose of carrying out a separate case study for the 7 DOF system is to effectively recognize the

utility of the RCCA method against similar conditions of damage in real time.

6.6.5 A note on the effect of the increased DOF on the detection results using the proposed algorithm

The central idea to the development of this section is to investigate the detection potential of the proposed RCCA algorithm towards an increased number of DOF. The 7 DOF B-W system is simulated based on the consideration that the two newly added floors exhibit a linear storey stiffness, that could be altered to induce a local damage to the system. The global damage cases, however, adhere to the previous notion of the change of the nonlinear force parameter at the base of the model, thereby imparting a temporal damage to the model. The detection results using the proposed algorithm have already been presented in the preceding sections and are not repeated here for brevity. Considering the Figs. 6.2 and 6.14, it can be very well established that the TVAR plots provide information regarding the exact damage instant to the system. However, the RRE plots presented in Fig. 6.14 provides an approximate estimation of the damage instant. As previously mentioned, the propagation of the change of the nonlinearity throughout the model is comparatively reduced due to an increased DOF of the system. Therefore, although the effect of global damage is appropriately captured by the TVAR coefficients for both the cases, the RRE plots provide a rough estimation for the 7 DOF B-W system, which can be considered as a direct implication of the increased number of DOF.

The local damage cases, however, are expected to show similar results for both the systems in context. A summary of the detection results for the systems can be obtained from Figs. 6.4 and 6.15 for a 15% change in the linear storey stiffness. As the proposed method successfully provided consistent detection results for a 10% saptio-temporal damage for the 5 DOF B-W system (Fig. 6.5), it is expected that the method should present similar results when the model is replaced with an increased number of linear DOF, depicted in Fig. 6.16. It can therefore be safely concluded that due to the confinement of the damage to a single storey, the proposed method provides similar spatio-temporal damage detection results for both the aforesaid models, in real time.

6.6.6 Case study for a strongly nonlinear system: A 5 DOF structure modeled with Duffing oscillator on its 3rd floor

In this section, the efficacy of the proposed method is assessed against a strongly nonlinear 5 DOF system, modeled with a Duffing oscillator on its 3rd floor. While the source of nonlinearity is provided by the Duffing parameter α, obtained from the governing differential equation of motion for the system, the other 4 floors are idealized as having linear storey stiffness. The concept of global damage emanates from the previous notion of changing the nonlinear force term that brings about a state of temporal damage throughout the system and the local damage is numerically simulated by altering the linear storey stiffness of a floor at a particular instant of time. The system is excited using Gaussian white noise for a duration of 50s, sampled at 100 Hz. The cases of global and local damage are separately discussed next in detail.

The numerical simulation of global damage is carried out by changing the value of the nonlinear force parameter α at 31s, presented in two individual case studies of 20% and 15% change. In the later stages of the chapter, the performance evaluation of RPCA is also provided that inspects its potential to identify damage for a strongly nonlinear system. For the first case, TVAR modeling is adopted on the transformed responses to yield the TVAR coefficients, clearly shown in Fig. 6.17. The sharp peak of the TVAR coefficients, accompanied by a significant distortion in the temporal RRE plots confirms the exact instant of damage at 31s. On reducing the α value by a fine level of 15% in real time, the generated recursive DSF plots are shown in Fig. 6.18. It can be clearly understood from the figure that the proposed algorithm is able to effectively identify the instant of damage for a fine level as low as 15%, which is a tough accomplishment considering the strongly nonlinear nature of the system and the recursive essence of the proposed method. It can be envisaged that the performance of the RPCA algorithm in evaluating damage for a strongly nonlinear system might not show consistent results, based on the premise that the RPCA methodology is pivoted around identifying damage patterns of linear to weakly nonlinear systems. Nevertheless, a fare comparison of these methods in the subsequent sections is attempted.

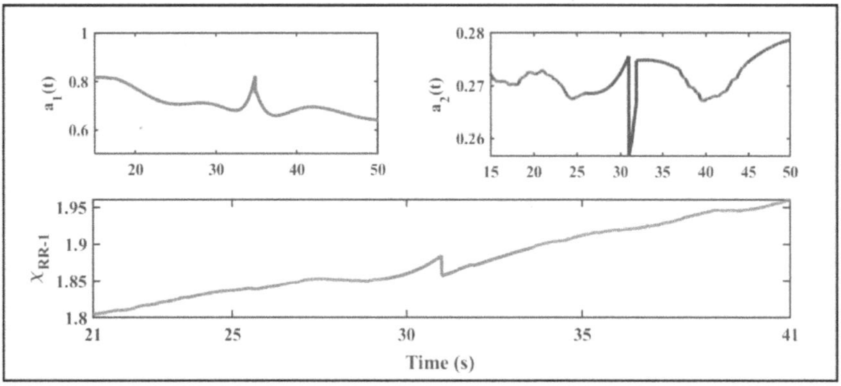

Figure 6.17: Detection results for 20% temporal damage for the strongly nonlinear system.

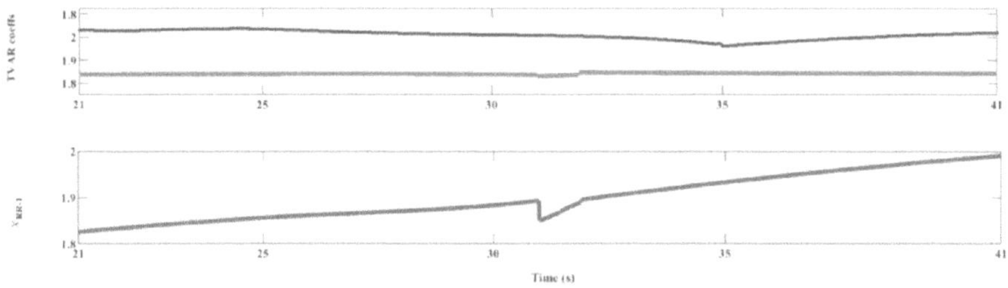

Figure 6.18: Detection results for 15% temporal damage for the strongly nonlinear system.

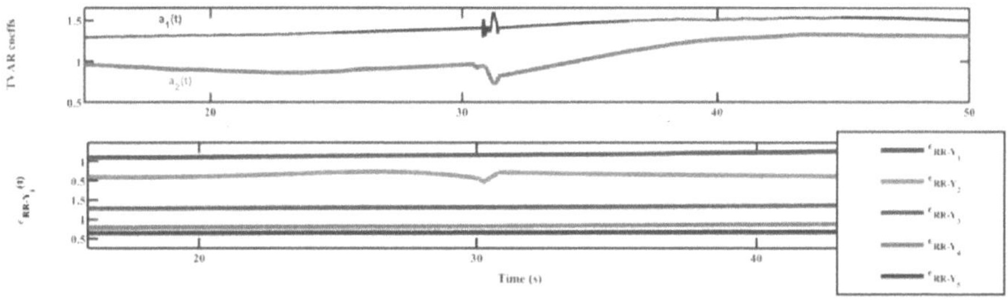

Figure 6.19: Spatio-temporal damage detection for strongly nonlinear system for 20% damage.

Proceeding to the next case, the spatio-temporal damage detection prowess of the proposed method is examined in a simultaneous framework. conforming to the previous notion of change in linear storey stiffness as an indicator of local damage, the 2nd floor of the model undergoes a stiffness change at 31s from the start. The damage cases of 20% and 15% are discussed next in detail. The detection results for the first case are displayed in Fig. 6.19, from which the exact instant of damage at 31s is clearly visible through the TVAR plots. Once the temporal damage is ascertained, the RCCA algorithm shifts to the spatial module where the local RREs for all the floors are tracked recursively. A significant distortion in the 2nd floor RRE and its absence in others, confirms the location of the damage, confined to the 2nd storey. Similar to these findings, the 15% spatio-temporal damage detection case also provides adequate evidence to suggest the instant of damage at 31s and its localization in the 2nd storey of the system. Considering the strongly nonlinear behavior of the system and the simultaneous spatio-temporal damage detection capability of the proposed method, the above findings hold a major entitlement of this work.

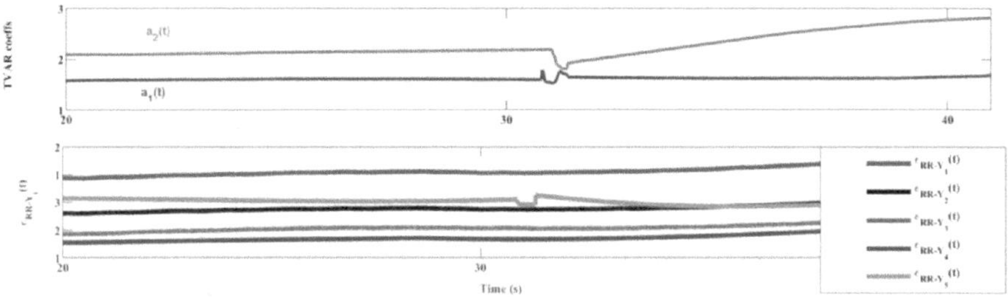

Figure 6.20: Spatio-temporal damage detection for strongly nonlinear system for 15% damage.

6.6.7 A case study with negligible non-linearity: Spatial damage detection

In this section, the potential of the proposed RCCA towards an almost linear system is examined. The previous discussions were mainly centered around non-linear systems and therefore, it becomes imperative to understand the applicability of the method towards a linear system as well. Obtaining a purely linear system is impractical, considering the material properties and the geometric nature of the system; hence, the nonlinear parameter α of the above mentioned model is scaled down to 0.005, which is negligible and can closely emulate to a linear state of the system. The spatio-temporal damage detection case is studied by considering a 20% damage to the system at 31s from the start, to the 2nd storey of the model. It can be clearly observed from Fig. 6.21 that the TVAR plots are effective at identifying the temporal damage instant through a sharp peak at 31s. Once the spatial module gets invoked, the local RREs for the 2nd floor show a significant distortion that confirms the presence of damage confined to that storey. Hence, it can be concluded that the proposed method is effective in identifying the spatial and temporal patterns of damage for both linear and nonlinear systems to a fair degree of accuracy, as evident from the underlying findings.

6.7 Performance check of the proposed method against RPCA: A comparative study

In this section, the performance check of the proposed method against a recently established real time damage detection scheme, utilizing the concepts of RPCA, is provided. The performance evaluation of the proposed method against its recently established counterpart takes place in three major case studies:

1. Global and local damage detection case studies for the 5 DOF B-W system excited using white noise.

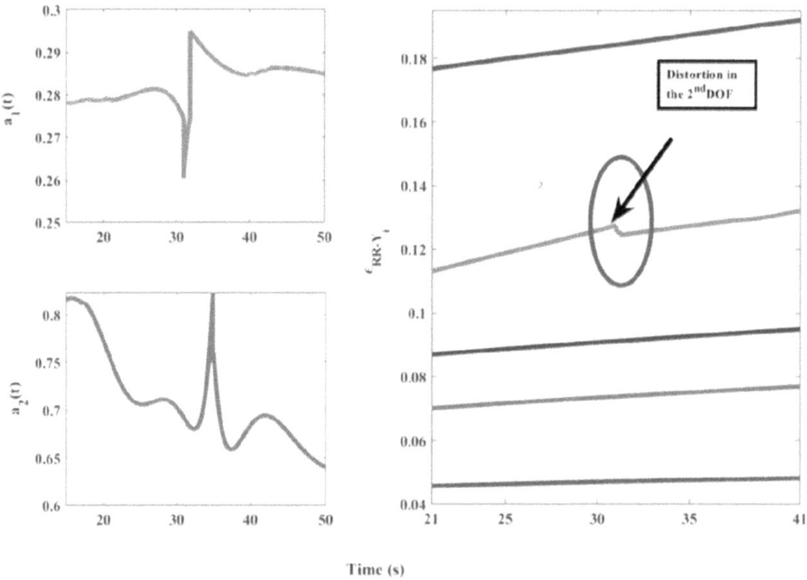

Figure 6.21: Spatial damage detection for almost negligible non-linearity (α = 0.05).

2. Global damage detection studies for the 5 DOF B-W system excited using El Centro ground motion.

3. Local damage detection studies for the 5 DOF system modeled using a Duffing oscillator on the 3rd floor, excited using white noise.

6.7.1 Case studies for the 5 DOF B-W system using white noise

The 5 DOF B-W system previously described is taken into consideration to examine the real time damage detection performance of both RCCA and the RPCA algorithms. The RPCA algorithm is a recently established real time scheme that has been published in the Journal of Mechanical Systems and Signal Processing, and is an important contribution in the context of real time damage identification of structures. The basic premise of the method was centered around finding damage for linear to weakly nonlinear systems. Since RPCA and RCCA both consider the EVD of the covariance matrices through their formulations, it becomes imperative to understand the performance of both the said methodologies for the same damage conditions. In this context, the global damage cases are first presented, followed by the results of the local damage detection.

6.7.1.1 Case study for 15% global damage detection using 5 DOF B-W system

The global damage is carried out by changing the nonlinear force term κ at 31s by 15%. Detection results obtained using RCCA can be found in Fig. 6.4. The RPCA algorithm operates in a similar fashion by first creating a covariance matrix out of the physical responses, performing a batch PCA on the initial samples and then using FOEP to estimate the eigenspace updates at each instant of time. Using TVAR coefficients, the transformed responses are modeled and the detection results using both the algorithms are shown Fig. 6.22. It is evident from the plot that the RPCA algorithm shows certain instabilities for such a low percentage of damage in real time, which can be primarily attributed due to the nature of the covariance matrix obtained from the physical responses. In comparison, the RCCA method creates a block covariance matrix that provides eigenspace updates at each time instant, and is therefore, devoid of instabilities of any kind. Moreover, the RPCA algorithm is premised on mostly linear system theory and its applicability towards a weakly nonlinear system is its direct extension. Therefore, it can be concluded that the RCCA method provides better estimates of detectability even for a fine level of damage in real time as compared to the RPCA scheme. However, it should be noted that both the algorithms provide a reasonable first-hand representation of the instant of damage in real time.

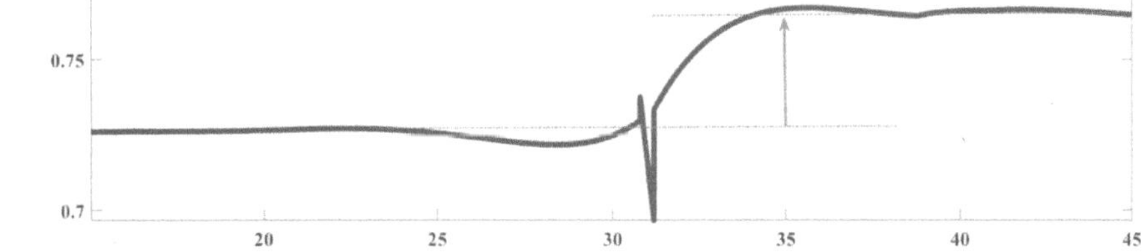

Figure 6.22: Comparison between RCCA and RPCA for 15% global damage using 5 DOF B-W system (under white noise excitation).

6.7.1.2 Case study for 25% local damage detection using 5 DOF B-W system

In this section, the performance of the FOEP based RPCA and RCCA methods are examined for a 25% spatio-temporal damage. A snapshot of the comparative study is provided in Fig. 6.23. It can be observed from the plot that both the recursive algorithms are able to identify the exact instant of damage at 31s that occurs in the 3rd storey of the system. However, from a straightforward visual inspection of the figure, it can be understood that the RPCA algorithm is susceptible to certain instabilities that might arise due to the inherent nonlinear nature of the system. Contrary to this, the RCCA method provides smooth detection results and therefore, preferred in identifying damage for a nonlinear system over the recently established RPCA framework.

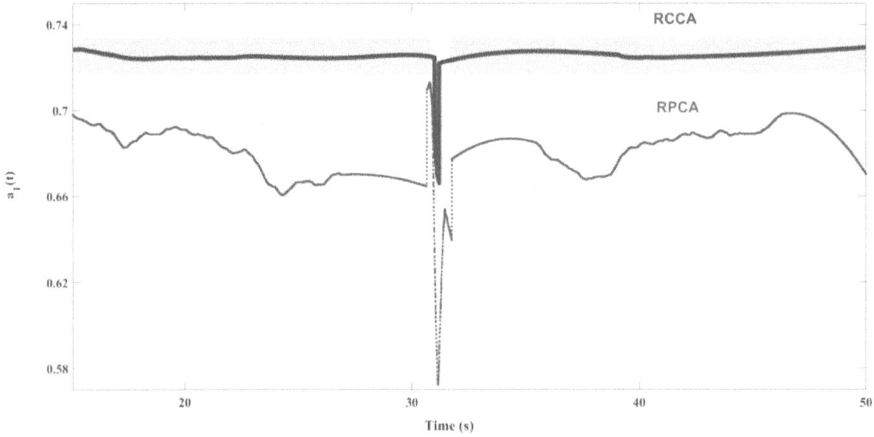

Figure 6.23: Comparison between RCCA and RPCA for 25% local damage using 5 DOF B-W system (under white noise excitation).

6.7.2 Case study for the 5 DOF B-W system excited using El Centro ground motion

A review of the recently established real time damage detection literature reveals that a 30% global damage for a nonstationary excitation is reported as the lower limit of detectability. However, the present work has successfully proposed *spatio-temporal* damage detection studies as fine as 20%, in real time. Based on these discussions, an even comparison with the RPCA method can only be justified at its lower limit of detectability, i.e., at 30% global damage. The detection results provided in Fig. 6.24 show a clear instant of damage at 31s using both the

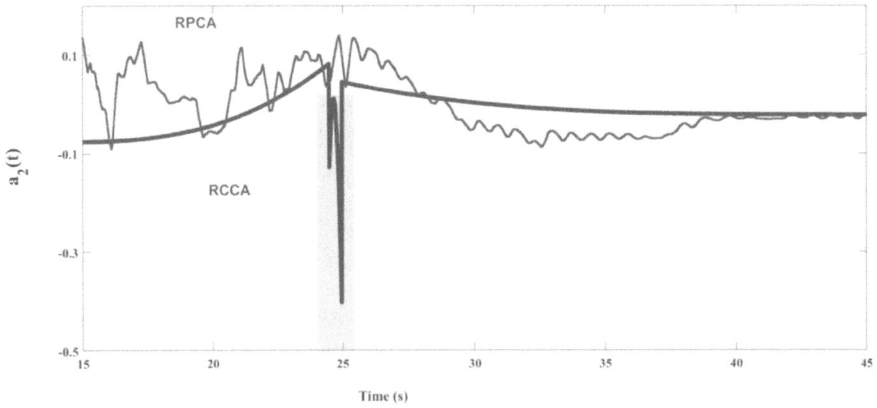

Figure 6.24: Comparison between RCCA and RPCA for 30% global damage using 5 DOF B-W system (under El Centro excitation).

algorithms. While the TVAR plots using RCCA provide impeccable detection results, the plots obtained using RPCA are highly susceptible to the nonstationarities that arise due to the input excitation. This creates a zone of significant instability in the TVAR plot, thereby rendering the application of RPCA under a nonstationary environment, ineffective. Hence, it can be safely concluded that the proposed RCCA algorithm provides better damage detection results in real time even for a nonstationary excitation for a nonlinear system as well.

The percentage changes in the mean level of the TVAR coefficients are shown in Table 6.2. It is well understood that the wavy modulations indicated by the fluctuations in the TVAR plots obtained from the RPCA transformed responses can be accounted for through the recursive standard deviations (SD) pre and post damage. It can be inferred from the statistical descriptions shown in Table 6.2 that the RPCA algorithm provides a significantly higher percentage change of the recursive SD before and after damage. This can mostly be attributed due

Table 6.2: Percentage changes in statistical mean of TVAR coefficients.

Damage condition	Method	Mean (undamaged)	Mean (damaged)	% change	Mean+SD (undamaged)	Mean+SD (damaged)	% change
15% global	RCCA	0.71	0.76	6.72	0.78	0.77	0.85
(WN)	RPCA	0.79	0.85	7.70	0.85	0.86	1.06
25% local	RCCA	0.68	0.69	1.47	0.73	0.70	4.92
(WN)	RPCA	0.67	0.69	2.98	0.72	0.69	4.98
30% global	RCCA	-0.01	-0.008	20	-0.009	-0.005	45.56
(El Centro)	RPCA	0.13	0.005	64.54	0.14	0.051	62.22

to the inconsistent modulations in the TVAR plots obtained from the RPCA algorithm, clearly depicted in Figs. 6.22, 6.23 and 6.24. A change in the mean level of the TVAR coefficients obtained from both the algorithms correspond to the occurrence of damage inflicted to the system at 31s. The higher percentage change of the statistical parameters reflected by the RPCA-TVAR algorithm characterize the presence of instabilities in the recursive DSF plots.

6.8 Experimental verifications

A crucial feature of any algorithm is to assess its effectiveness by validation through experimental case studies, carried out under a controlled laboratory environment. In the recent years, numerous experimental setups have been developed that closely emulate real time damage scenarios. In the present work, the focus has shifted towards detecting the change of state from a linear to nonlinear regime and vice versa, through an extensive experimental setup. To substantiate the robustness of the proposed approach, an experimental setup is devised consisting of a two-storey aluminum shear building prototype placed with a water dispenser at its top floor. The aluminum model is of height $1.5m$ with an inter-storey distance of $0.74m$ fixed on a base plate that is bolted on top of a shake table. The cross section of the base measures $0.3m \times 0.3m \times 0.02m$ placed on the $0.15m$ thick base plate and firmly bolted against the shake table. The water dispenser of height $0.26m$ is firmly affixed to the roof plate using commercially available epoxy resin. The model is subjected to scaled Gaussian white noise excitation and the acceleration data is collected using a QuantumX MX410 HBM Data Acquisition (DAQ) system at a sampling frequency of $400Hz$. The shear building model is instrumented at the floor levels in orthogonal directions using PCB Piezotronics triaxial acceleormeters of sensitivity $100mV/g$, bearing a frequency range of $0.5 - 10000Hz$. The dispenser when filled with water provides nonlinearity to the setup. This comes into effect when the water sloshes against the walls of the dispenser as the model is subjected to scaled vibration excitation. The flow valve of the dispenser emancipates water at the rate of $1L/30s$ during the run-time of the experiment.

The experimental trials were conducted at two different alignments of the shear building model (Fig. 6.25): **Alignment A**: The model is placed along the axis of the shake table with the local x-axis coinciding with the global x-direction and **Alignment B**: The model positioned at a non-orthogonal orientation with respect to the global x- and y-axes. For each of these cases, the experiment is carried out by allowing the water to drain out of the dispenser at 0s and at 20s in separate trials, that continued for the remainder of the excitation. The time stamp of the trial was captured using a stop watch and the experiment was also recorded in the form of videos, in order to gain an immaculate insight about the dynamics

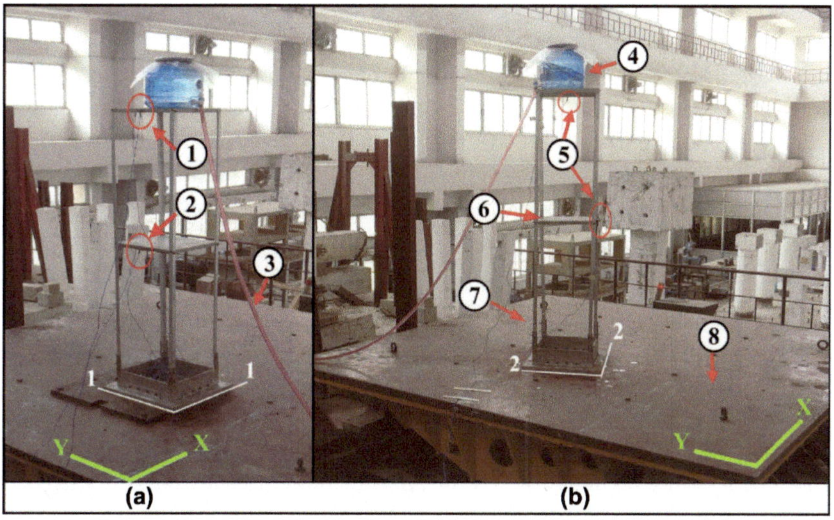

Figure 6.25: Details of the experimental setup: (a) biaxial and (b) orthogonal directions.

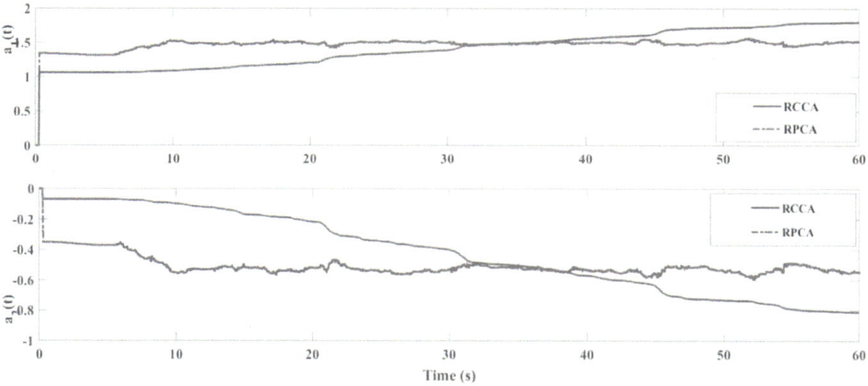

Figure 6.26: DSFs for mass loss at 0s, Alignment A.

of the process. The streaming acceleration data is recorded and the proposed algorithm is applied in real time that yields the transformed responses. As the water drains out from the system, the sloshing of the water leads to a variable mass loss that accounts for the transition from nonlinear to linear state, which can be distinctly identified using the proposed method, described next in detail.

In addition to demonstrating the detection potential of the RCCA algorithm, the performance of the recently established RPCA method is also dealt with in detail. Based on the previous discussions, it can be easily understood that the RPCA algorithm premises on the concept of orthogonal transformations for data gathered from linear to weakly nonlinear systems. Its performance for a multi-directional case can therefore, be expected to falter as the formulation of the algorithm does not take into consideration the effects of torsionally coupled systems involving multi-directional dependencies. In this regard, the application of both the algorithms on the streaming dataset from the experimental trials generates eigenspace updates at each time instant, from which the transformed responses are modeled using the TVAR approach. The comparative study of the performance of these recursive algorithms is discussed in the following section.

6.8.1 Detection results for the experimental trials: A comparative study with RPCA

The behavior of the recursive DSFs for Alignment A are illustrated in Fig. 6.26 and 6.27, for mass loss at 0s and 20s, respectively. For the case where the water drains out at 0s from the start, Fig. 6.26 shows a gradual change in the mean level of the TVAR coefficients. The draining of water is a continuous event, which necessitates that the recursive damage markers should be able to illustrate a similitude through their representations. This conceptualization can be exactly inferred from the TVAR coefficients where consistent changes in the mean level of the plots correspond to a gradual loss of water from the system. This clearly depicts the transition from a nonlinear to a linear state of the model, validated by the DSF plots. Considering the TVAR plots obtained from the RPCA algorithm, the coefficients displayed in Fig. 6.26 indicate the initial regime, but the inconsistency of the plots gradually impedes its expected utility. The loss in mass due to water drainage is a gradual event, a fact corroborated by the DSFs obtained from RCCA transformed response. In the present case, the RPCA algorithm fails to provide any substantial evidence of the gradual mass loss scheme, conducted through the experiment. However, the global change in the mean level from the 0s mark until the entire experimental duration is indicative of a variable mass loss and can be roughly interpreted as a transition of state of the system. In line with the above findings, the DSFs shown in Fig. 6.27 clearly identify the commencement of water loss from 20s from the start of the excitation. It can be inferred that the water loss from 20s represents a gradual event, evident from the consistent shifts in the mean level of the TVAR plot. An interesting feature to observe here is that the graph maintains an apparent horizontal alignment until the commencement of water loss at 20s, illustrating the fact that the transition from the nonlinear to the linear regime is aptly depicted by the proposed algorithm. The detection results using RPCA are not expected to show consistent changes indicating the phenomena of the mass loss, the results of which are omitted here for brevity.

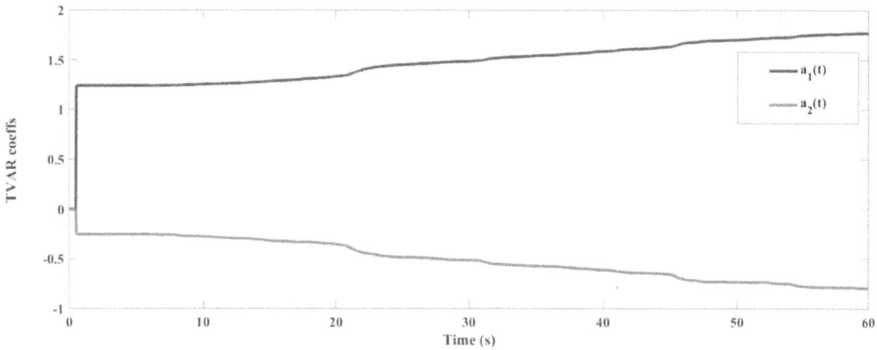

Figure 6.27: DSFs for mass loss at 20s, Alignment A.

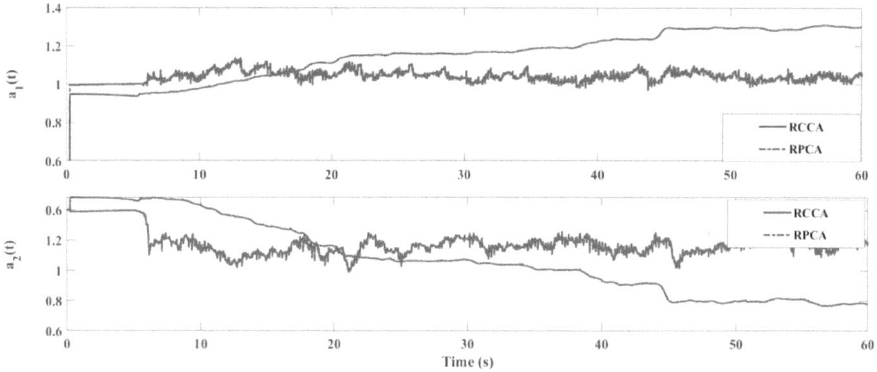

Figure 6.28: DSFs for mass loss at 0s, Alignment B.

The subsequent trials on the experimental setup were conducted by aligning the model in a non-orthogonal orientation with respect to the shake table (Alignment B). A similar procedure is adopted where the variable mass loss takes place at 0s and 20s, for two individual case studies. The main objective of this experimental trial is to assess the efficacy of the proposed algorithm for a multi-directional case, closely emulating a practical scenario, where the ground motions might excite a structure from a non-orthogonal alignment. Motivated by this objective, the RCCA method is applied on the biaxial (non-orthogonal) datasets obtained from the streaming vibration response. From the experimental trials, it was observed that the sloshing of water due to the excitation in the aligned direction had a greater magnitude compared to the mutually orthogonal oriented cases. Figure 6.28 clearly shows that the change in the mean level of the DSF plot takes place from 0s, thereby conforming to the fact that the water loss

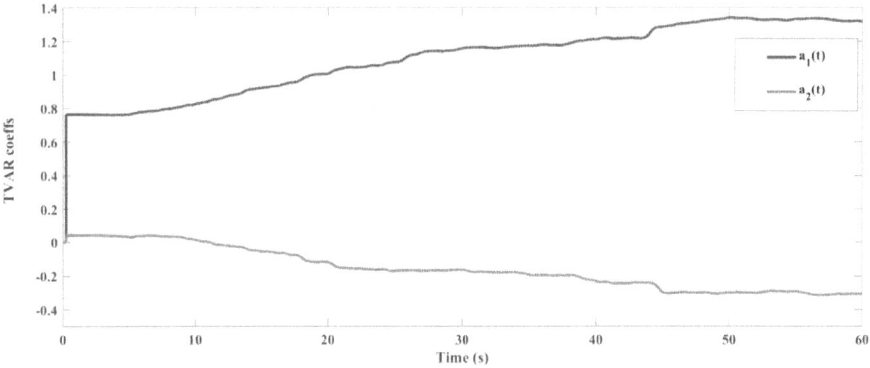

Figure 6.29: DSFs for mass loss at 20s, biaxial orientation.

commences exactly at the beginning of the excitation. However, it is observed that the shifts in the mean level of the TVAR coefficients is not uniform, which can be attributed to the multi-directional nature of the recorded responses. The torsional component generated during the vibration of the shake table allowed a rapid sloshing of water with a greater magnitude, compared to the previous case. It can be clearly observed from Fig. 6.28 that the TVAR coefficients obtained using the RPCA algorithm fail to show consistent deviation in the mean level, an occurrence that had immaculately been corroborated by the RCCA algorithm. Furthermore, the gradual transition of state from the 0s mark is not obvious, attributed primarily to the global non-orthogonal pairs of vibration inputs. The comparative study with the recently established RPCA method conclusively indicates that the RCCA algorithm provides better detection results even when the input features are locally orthogonal with respect to each other.

A similar case study for the DSFs involving water loss at 20s is reported in Fig. 6.29. While the change in the mean level of the plots is pronounced from 20s, the shifts in the level is not consistent. It can therefore, be concluded that the proposed RCCA based approach provides better identification results when the vibration data streams in from a pair of globally orthogonal sensors, as opposed to the ones obtained from the pair of multi-directional locally orthogonal datasets.

References

[1] Kerschen, G., Golinval, J. C., Vakakis, A. F. and Bergman, L. A. (2005). The method of proper orthogonal decomposition for dynamical characterization and order reduction of mechanical systems: An overview. Nonlinear Dynamics, Springer, 41(1-3): 147–169.

[2] Thompson, B. (2000). Canonical Correlation Analysis.

[3] Härdle, W. K. and Simar, L. (2015). Canonical correlation analysis. Applied Multi-variate Statistical Analysis. Springer, pp. 443–454.

[4] Fan, X. and Konold, T. R. (2010). Canonical correlation analysis. Quantitative Methods in the Social and Behavioral Sciences: A Guide for Researchers and Reviewers, pp. 29–34.

[5] Thorndike, R. M. (2000). Canonical correlation analysis. Handbook of Applied Multi-variate Statistics and Mathematical Modeling. Elsevier, pp. 237–263.

[6] Akaike, H. (1976). Canonical correlation analysis of time series and the use of an information criterion. Mathematics in Science and Engineering, Elsevier, 126: 27–96.

[7] Thompson, B. (1984). Canonical Correlation Analysis: Uses and interpretation. No. 47. SAGE Publications.

[8] Statheropoulos, M., Vassiliadis, N. and Pappa, A. (1998). Principal component and canonical correlation analysis for examining air pollution and meteorological data. Atmospheric Environment, 32(6): 1087–1095.

[9] Bajorski, P. Canonical correlation analysis. Statistics for Imaging, Optics, and Photonics, pp. 241–259.

[10] Vía, J., Santamaria, I. and Pérez, J. (2005). A robust RLS algorithm for adaptive canonical correlation analysis. In Acoustics, Speech, and Signal Processing, 2005. Proceedings.(ICASSP'05). IEEE International Conference on, IEEE, 4: iv–365.

[11] Bach, F. R. and Jordan, M. I. (2005). A probabilistic interpretation of canonical correlation analysis.

[12] Ouarda, T.. B., Girard, C., Cavadias, G. S. and Bobée, B. (2001). Regional flood frequency estimation with canonical correlation analysis. Journal of Hydrology, 254(1-4): 157–173.

[13] Russell, E. L., Chiang, L. H. and Braatz, R. D. (2000). Fault detection in industrial processes using canonical variate analysis and dynamic principal component analysis. Chemometrics and Intelligent Laboratory Systems, 51(1): 81–93.

[14] Schólkopf, B., Smola, A. and Múller, K. R. (1998). Nonlinear component analysis as a kernel eigenvalue problem. Neural Computation, 10(5): 1299–1319.

[15] Feng, L., Yi, X., Zhu, D., Xie, X. and Wang, Y. (2015). Damage detection of metro tunnel structure through transmissibility function and cross correlation analysis using local excitation and measurement. Mechanical Systems and Signal Processing, Elsevier, 60: 59–74.

[16] Hazra, B. and Narasimhan, S. (2010). Wavelet-based blind identification of the UCLA Factor building using ambient and earthquake responses. Smart Materials and Structures, IOP Publishing, 19(2): 025005.

[17] Hazra, B., Sadhu, A., Roffel, A. J. and Narasimhan, S. (2012). Hybrid time-frequency blind source separation towards ambient system identification of structures. Computer-Aided Civil and Infrastructure Engineering, 27(5): 314–332.

[18] Musafere, F., Sadhu, A. and Liu, K. (2015). Towards damage detection using blind source separation integrated with time-varying auto-regressive modeling. Smart Materials and Structures, IOP Publishing, 25(1): 015013.

[19] Sadhu, A. and Hazra, B. (2013). A novel damage detection algorithm using time-series analysis-based blind source separation. Shock and Vibration, IOS Press, 20(3): 423–438.

[20] Pakrashi, V., Fitzgerald, P., OLeary, M., Jaksic, V., Ryan, K. et al. (2018). Assessment of structural nonlinearities employing extremes of dynamic responses. Journal of Vibration and Control, SAGE Publications, 24(1): 137–152.

[21] Cichocki, A. and Amari, S. I. (2002). Adaptive Blind Signal and Image Processing: Learning Algorithms and Applications. Vol. 1. John Wiley & Sons.

[22] Golub, G. H. and Zha, H. (1994). Perturbation analysis of the canonical correlations of matrix pairs. Linear Algebra and its Applications, 210: 3–28.

[23] Kato, T. (2013). Perturbation Theory for Linear Operators. Springer Science & Business Media, Vol. 132.

[24] Feeny, B. F., and Kappagantu, R. (1998). On the physical interpretation of proper orthogonal modes in vibrations. Journal of Sound and Vibration, Elsevier, 211(4): 607–616.

[25] Li, W., Yue, H. H., Valle-Cervantes, S. and Qin, S. J. (2000). Recursive PCA for adaptive process monitoring. Journal of Process Control, Elsevier, 10(5): 471–486.

Chapter 7

Real-time Modal Identification—RCCA Based Approach

A reference free approach for identifying structural modal parameters using recursive canonical correlation analysis (RCCA) will be discussed in this chapter. This discussion is a natural extension of the applicability of FOEP based RCCA towards identification of vibratory modes for linear systems in a recursive framework. The method provides eigenspace updates at each instant of time that provide linear normal modes (LNMs) of a vibrating system. Modelled in a recursive framework, a traditional measure of mode-correlation – the modal assurance criterion (MAC) – is examined in a recursive framework. This provides a direct comparison between the theoretical modes against the estimated LNMs which eventually verifies the similitude in the obtained modes. Moreover, these eigenspace updates provide basis vectors that can be subsequently utilized to determine potential damage induced to the system. The set of transformed responses obtained at each time stamp is fit using time varying auto regressive (TVAR) models in order to aid as damage sensitive features (DSFs) for identifying temporal and spatial patterns of damage. In the preceding chapter it was discussed that the RCCA algorithm is adept at accommodating multi-directional dependencies of a system. Numerical simulations carried out on linear systems demonstrate the efficacy of the proposed method in recursively separating the structural modes. Comparative case studies with its batch counterpart – the traditional CCA based method – illustrates the exemplary performance of the RCCA scheme in identi-

Table 7.1: List of Acronyms.

CCA	Canonical Correlation Analysis	PCA	Principal Component Analysis
POC	Principal Orthogonal Component	POM	Proper Orthogonal Mode
EVD	Eigen Value Decomposition	BSS	Blind Source Separation
ICA	Independent Component Analysis	RCCA	Recursive Canonical Correlation Analysis
TVAR	Time Varying Auto Regressive	DSF	Damage Sensitive Features
RRE	Recursive Residual Error	DOF	Degree of Freedom
SHM	Structural Health Monitoring	FOEP	First Order Eigen Perturbation

fying the vibratory modes of a system in real-time. It is this important aspect that provides asset managers and infrastructure owners accurate information about the state of the monitored system, which is greatly missing in the context of appropriate industrial practice. Embedded in the same framework is the applicability of the proposed approach towards detecting spatio-temporal structural damage in real-time for both dynamic systems. The chapter also provides details about experimental trials carried out under controlled laboratory environment aimed at accurately preserving structural modal information and detecting changes of state for a test setup. The results obtained features an assertive evidence-base suggesting the use of RCCA as a potent candidate for real-time modal identification and damage detection in an integrated framework.

7.1 Motivation

Traditional structural modal identification [1–7] involves a wide class of exploratory methods that estimate modal properties through statistical methods in both time and frequency domain (and in some cases, time-frequency methods) [8]. Most of the parametric identification techniques – like natural excitation technique (NExT) [9], stochastic subspace identification (SSI) [10] and eigen system realization (ERA) [11] – pose serious problems of model order selection, that become more involved in presence of spurious modes. In retrospect, the technical forte to continuously monitor infrastructures is hindered due to the offline implementation of such algorithms [12, 39–41]. Algorithms based on arbitrary windowing but not mathematically consistent with the paradigm of continuous monitoring, have limited SHM applications [13–15]. Especially for time-varying systems – such as tuned mass dampers (TMDs) which are frequently prone to mistuning [16], wind turbines [17], and systems undergoing bridge-vehicle interaction [18] – substantial information is lost in offline analysis, data transmission, and pre-processing [19]. To alleviate this drawback, modal properties should be extracted in real-time to elicit continuous-time modal information. Recently developed online damage detection algorithms [13–15] ensure capabilities of providing real time estimates of proper orthogonal modes (POMs). A natural extension of this concept can effectively cater to real-time modal identification that aims at providing online estimates of linear normal modes (LNMs) of a vibratory systems (which essentially correspond to the theoretical modes). In this context, an exploratory class of algorithms investigating the covariance structure of the physical responses in multiple independent blocks (instead of multiple *time lags*) is developed for identifying LNMs. This chapter provides a comprehensive treatment of *recursive canonical correlation analysis* (RCCA) as a candidate for real-time modal identification and compares it against other alternatives for various numerical, experimental and benchmark examples.

Recent studies in statistical signal processing has exemplified source separation strategies – most common of them include blind source separation (BSS) [1, 2, 6, 16, 20, 21] and independent component analysis (ICA) [8, 22, 23] based techniques. It is interesting to notice how algorithms centred around general eigenvalue decomposition (GEVD) and standard EVD, joint diagonalization, unitary factorization, and simultaneous diagonalization of the covariance matrix of the measured physical response, have played a major part in today's SHM practices worldwide. Notably, algorithms such as principal component analysis (PCA) [24, 25] and BSS [1, 2, 20–22, 26, 27], have significantly contributed to a exhaustive application of structural dynamics problems. A sophisticated opinion on equivariant adaptive separation via independence approach (EASI) [12], has promised hybrid modal identification in recent times, that basically involves a windowing approach to identifying the vibratory parameters of a system. Although the algorithm provides a certain degree of improvement over its traditional counterparts, mathematical formulations implicitly adopting ICA bases leaves room for studying a host of other BSS algorithms that investigates the covariance structure of real-time response signals. The work carried out in this chapter is motivated by: (i) the potential use of CCA as a structural modal identification tool and (ii) its natural extension towards RCCA that automatically embeds FOEP based LNM extraction in real-time, thereby facilitating online modal identification.

A powerful statistical technique developed in the recent years, CCA, utilizes the *block* covariance matrix of the physical responses in order to measure the underlying correlation between two sets of multidimensional variables [28–30]. CCA attempts to find the projections in an optimal subspace so that the mutual correlation coefficient between a pair of variables is maximized [30]. Contrary to the traditional methods such as second order blind identification (SOBI), parallel factor decomposition (PARAFAC) and ICA, CCA yields block covariance eigenspace that accommodates for a multi-directional application in vibrating systems. As the family of BSS methods come under the category of CCA based schemes, the direct application of CCA towards modal identification is eventually justified. However, it should be noted that most of the OMA methods are offline in nature that essentially requires batches of windowed data to operate [8, 12, 26]. In this context, BSS algorithms belonging to the family of joint diagonalization [5, 16, 20] and non-Gaussianity maximization, are not readily amenable towards recursive formulation in the context of modal identification [16,20,21,27]. Therefore, for continuous condition-based monitoring of systems, algorithms must work *online* and provide real time updates, an aspect that has not been attempted in the context of CCA based algorithms. In order to implement a real time identification approach, the use of FOEP techniques provide rank one updates of the eigenspace at each instant of time, from which the modal parameters can be determined [13–15, 31, 32]. One of the primary advantages of an online monitoring framework lies in the processing of streaming data, as opposed to the traditional identification approaches. The availability of a mathematically consistent real time framework relieves the recourse to ad-hoc windowing of the data. Since real time modal identification involves online update of mode shapes as well as modal responses, this information could be possibly utilized to perform *real time damage detection* without any extra post processing. Although these problems are mutually exclusive (i.e., each can be individually addressed), but a single framework that can adequately address both these aspects without taking recourse to extra post processing, can potentially develop into a robust SHM system. Real time damage detection is thus considered in the present work as an application example to illustrate efficacy and utility of the proposed RCCA based real time modal identification algorithm.

This chapter discusses the development of RCCA as a stand-alone online vibration based SHM algorithm that identifies modal parameters in real time for continuously streaming data. A survey of the literature reveals that real time strategies aimed at detecting damage online fails to perform well for cases where multi-directional dependencies are considered [13, 14]. In this regard, RCCA accommodates the feature space arising from multiple sources of excitation locally oriented against each other to perform multivariate statistical analysis, in real time. Additionally, the recursive eigenspace updates obtained is further shown to provide information regarding the spatio-temporal patterns of damage, pitched in as an application specific problem. A unified framework incorporating the aspects of system identification and damage detection in a recursive framework has not been attempted in the associated literature. With the improvement in the level of detection in real time, understanding the performance of the systems subjected to stochastically modeled load becomes important. This will contribute to an extensive and informed condition based monitoring process in the context of damage diagnosis and prognosis strategies.

7.2 Background

The previous chapter describes CCA as an exploratory method for determining the relationship between two multivariate datasets through their *auto and cross-covariance* matrices. Recall that CCA serves as a standard tool in statistical analysis that finds its application in econometrics, meteorology, medical studies and signal processing [28,29]. The present work explicitly details the preliminary attempt at integrating CCA into a real time, structural modal identification framework. As previously mentioned, the algorithm finds two basis vectors in which the correlation matrix between the variables is diagonal and mutually maximized by projecting the original set of variables onto an optimum subspace [30]. The BSS based consideration of CCA provides the justification of the approach towards modal identification as outlined below.

7.2.1 BSS based formulation of CCA

The use of BSS in dynamic system identification has augmented in recent years, evident from the deluge of publications in this area [1,5,6,26]. Imposing certain conditions on the input spectral characteristics, BSS assumes the normal modes of a linear dynamical system to correspond to *virtual sources* or *virtual inputs*. Therefore, separating these virtual inputs from a mixture of sources (essentially signals) using only the information contained in a set of acquired physical response naturally maps to a problem of modal identification by itself [1,16,20,27]. Hence, the term *blind* is commonly contextualized. It is interesting to note how BSS – a source separation algorithm – can be transformed (or 'mould') into a cast of inverse structural dynamics problems. This requires a basic understanding of the covariance matrix of random variables structured in the form of a block matrix:

$$\mathbf{C} = \begin{bmatrix} \mathbf{C}_{uu} & \mathbf{C}_{uv} \\ \mathbf{C}_{vu} & \mathbf{C}_{vv} \end{bmatrix} \tag{7.1}$$

where $\mathcal{U}$ and $\mathcal{V}$ are the corresponding linear projections of two zero centered datasets U and V. Note that the theoretical development of CCA (and subsequently, *RCCA*) discussed here is purely in the light of BSS based approaches. The construct in the previous chapter sketched CCA in terms of a signal processing statistical approach and subsequently identified structural damage detection as a viable problem to address. The key difference here is the transformation of CCA as a BSS based modal ID strategy that has not been previously approached in the literature. Using indigenously developed FOEP approaches, the mapping of CCA to RCCA will become evident in the following discussions.

The aim of this methodology is to obtain the maximized canonical correlations in the optimum subspace between $\mathcal{U}$ and $\mathcal{V}$, that can be easily obtained by solving the following sets of eigen decomposition equations:

$$\left.\begin{aligned} \mathbf{C}_{uu}^{-1}\mathbf{C}_{uv}\mathbf{C}_{vv}^{-1}\mathbf{C}_{vu}\hat{\mathbf{w}}_u = \rho^2\hat{\mathbf{w}}_u \\ \mathbf{C}_{vv}^{-1}\mathbf{C}_{vu}\mathbf{C}_{uu}^{-1}\mathbf{C}_{uv}\hat{\mathbf{w}}_v = \rho^2\hat{\mathbf{w}}_v \end{aligned}\right\} \tag{7.2}$$

where the eigenvalues ρ^2 represent the squared *canonical correlations* and the corresponding eigenvectors $\hat{\mathbf{w}}_u$ and $\hat{\mathbf{w}}_v$ denote the normalized the canonical correlation *basis vectors*, respectively. In engineering applications, a matrix form of Eq. 7.2 is usually approved of, thereby recasting it into a single general eigenvalue equation, given by:

$$\mathbf{B}^{-1}\mathbf{A}\hat{\mathbf{w}} = \rho\hat{\mathbf{w}} \tag{7.3}$$

where the matrices involved in the eigenvalue decomposition (EVD) are obtained as:

$$\mathbf{A} = \begin{bmatrix} \mathbf{0} & \mathbf{C}_{uv} \\ \mathbf{C}_{vu} & \mathbf{0} \end{bmatrix}, \quad \mathbf{B} = \begin{bmatrix} \mathbf{C}_{uu} & \mathbf{0} \\ \mathbf{0} & \mathbf{C}_{vv} \end{bmatrix} \quad and \quad \hat{\mathbf{w}} = \begin{pmatrix} \mu_u\hat{\mathbf{w}}_u \\ \mu_v\hat{\mathbf{w}}_v \end{pmatrix} \tag{7.4}$$

BSS, essentially being a source separation algorithm automatically embeds CCA in its formulation. CCA – primarily dependent on block covariance matrices – derive a similar structure to PCA. The special case where CCA corresponds to PCA considers the matrices **A** and **B** structured as: $\mathbf{A} = \mathbf{C}_{xx} \quad and \quad \mathbf{B} = \mathbf{I}$. Taking these inter-relations into consideration, CCA can be thought of as an overarching framework that embeds PCA (and its variants, such as KPCA, NNPCA, among others) and at the same time incorporates BSS. Towards this, it is now important to identify PCA (as a subset of CCA) in terms of inverse structural dynamics problems in general, and modal identification, in particular.

7.2.1.1 PCA and structural dynamics

A class of conventional inverse structural vibration problems involving exploratory analysis of proper orthogonal components (POCs), nonlinearity detection and SHM is carried out using a family of orthogonal transformation based algorithms, popularly known as PCA [13, 14, 24, 25, 33]. Consider the individual elements of the covariance matrix of the modal displacement response ($\mathbf{R_Q}$), given by [13, 14]:

$$r_{ij}^{Q} = \int_{\tau=0}^{\infty}\int_{\theta=0}^{\infty} f_i(\tau)f_j(\theta)\left[\frac{1}{N}\sum_{i,j} h_i(t-\tau)h_j(t-\theta)\right]d\tau d\theta \tag{7.5}$$

The POCs (ψ) can be expressed as product of POMs (**W**) and the data vector (U), according to: $\psi_i(t) = \mathbf{W}^{\mathbf{T}}u_i(t)$. Consequently, it can be easily inferred that the POCs simplify to a sum of true linear normal coordinates (q) and an error term (ε). The elements of the covariance matrix of the POCs can be represented as:

$$r_{ij}^{\psi} = \frac{1}{N}\sum_{k=1}^{N}\psi_i(t_k)\psi_j(t_k) = \frac{1}{N}\sum_{k=1}^{N}[q_i(t_k)q_j(t_k) + \varepsilon q_j(t_k) + q_i(t_k)\varepsilon + \varepsilon^2] \tag{7.6}$$

For practical systems with low to moderate damping and finite sample size, it can be understood from Eq. (7.6) that POC provides a good approximation to the true linear modal component which deviates from each other as damping increases. Hence the POC covariance matrix $\mathbf{R_\Psi}$ is expected to show a diagonally dominant behavior in the limit as $N \to \infty$ and when the errors are low (i.e., low to moderate damping) [25]. It has been observed that PCA does not provide accurate estimation of the theoretical LNMs for a vibrating system [13, 14, 27]. This automatically implies that the identification of normal modes in real time is difficult within the framework of PCA.

In recent times, a similar class of algorithms utilizing the multi-lagged temporal correlation structure of the responses, referred to as SOBI, have shown promise in this area. SOBI uses PCA as a primary step, but extends the orthogonalized components further to unitary factorization that simultaneously diagonalizes the stacked covariance matrix [16]. It should be noted here that PCA provides spatially decorrelated datasets for zero time lag, while the temporal structure of the sources still remain intact for non-zero time lags [43]. Therefore, complete modal recovery is not possible. In conjunction with unitary factorization procedure at one or several non-zero time lags, the complete decorrelation of the sources takes place. The application of SOBI has shown to yield better estimates where it easily outperforms ICA for high levels of damping [20, 21]. Additionally, an alternate possibility using matrix factorization for a multi-block covariance structure instead of a time-lagged covariance structure can be accommodated through the use of CCA [30]. Although CCA has shown to be a powerful alternative of using SOBI, its utilization towards modal identification is yet to be studied. Therefore, the present work considers the extension of CCA in real time that can be addressed through the use of FOEP based techniques.

7.2.2 Recursive covariance estimation and FOEP

Recent research in signal processing techniques has identified CCA as a potential candidate towards structural dynamics problems. The key objective of this study is to develop a real time framework that could employ the concepts of RCCA into a structural modal identification module and extend its applicability towards online damage detection. Towards this, certain formulations in the CCA based approach need to be modified in order to tailor the methodology towards an online implementation. The recursive implementation of CCA could possibly be carried out using symmetric QR algorithm [34], Lanczos tridiagonalization [35], Kalman Filter (KF) updates [36] and the FOEP approach [13–15], to name a few. However, a recent review of the supporting SHM literature suggests that although the KF updates provides state recursion estimates at each iteration, the method is greatly limited by the restrictive requirements on *apriori* information of the model structure and process and measurement noise. Previous case studies have shown that the use of QR algorithms and the block wise Lanczos tridiagonalization algorithms require $2.68m^3$ and $2.5m^3$ multiplications respectively (m being the number of eigenvectors), thereby rendering these processes computationally expensive. Contrary to these methods is the *FOEP based theory* that is essentially parameter free and requires $2m^3$ multiplications at each eigenspace update [31,37] easily obtained through an EVD on the initial datasets, essential for stabilizing the algorithm. The eigenspace updates so formed at each time stamp, are crucial in estimating the LNMs of a vibrating system and can be subsequently utilized to infer the damage patterns inflicted to the structure over time (if any).

An instructive approach towards the basic understanding of FOEP involves an assumption of the eigenvalues of a perturbed matrix $\mathbf{C}+\mathbf{\Delta C}$ to be of the form $\mathbf{\Lambda}+\beta\beta^T$, i.e., the *rank-one update* of the matrix $\mathbf{\Lambda}$. Defining the following expressions:

$$\begin{aligned} \mathbf{C\Psi} &= \mathbf{\Lambda\Psi} \\ (\mathbf{C}+\mathbf{\Delta C})(\mathbf{\Psi}+\Delta_{\mathbf{\Psi}}) &= (\mathbf{\Psi}+\Delta_{\mathbf{\Psi}})(\mathbf{\Lambda}+\Delta_{\mathbf{\Lambda}}) \end{aligned} \tag{7.7}$$

where, $\Delta_{\mathbf{\Psi}}$ and $\Delta_{\mathbf{\Lambda}}$ are the perturbation matrices. The factorization of the matrix $\mathbf{C}$ into its corresponding canonical form establishes the relation, $\mathbf{C} = \mathbf{\Psi\Lambda\Psi}^T$ with $\mathbf{\Psi\Psi}^T = \mathbf{I}$. On ignoring the second order perturbation terms, the EVD of the perturbed matrix $\mathbf{\Delta C}$ can be expressed as:

$$\mathbf{\Delta C} = \beta\beta^T = \mathbf{\Psi\Lambda}\Delta_{\mathbf{\Psi}}{}^T + \mathbf{\Psi}\Delta_{\mathbf{\Lambda}}\mathbf{\Psi}^T + \Delta_{\mathbf{\Psi}}{}^T\mathbf{\Lambda\Psi}^T + O(\epsilon^3) \tag{7.8}$$

It is worth noting the fact that the aforementioned expressions crucial to the development of the FOEP approach requires no apriori assumptions on the nature of the dataset. Hence, the varied applications of the FOEP methods range from analyzing binary datasets, monitoring chemical processes and tackling structural damage detection problems as well; a facet that has been explored in detail by the authors in the recent past. Based on the above discussions, the detailed derivation for the RCCA approach can now be explored in detail.

7.3 RCCA: Detailed derivation

In order to identify modes online, it is essential to estimate the eigenspace recursively, thereby providing iterative updates at each instant of time. The major challenge towards tailoring the basic CCA towards an online implementation is to perform EVD on the block covariance matrix, which is particularly cumbersome and memory consuming. This can be alleviated through the use of the FOEP based approach which provides recursive updates of the eigenspace of the data at a particular instant [13–15]. An important aspect to be considered here is that the recursive approach updates the eigenspace characteristics at each time instant, instead of *updating the covariance matrix as a whole*, thereby reducing the time complexity of the recursive implementation [13–15].

The theoretical development of RCCA is premised on the objective of finding a recursive update of the eigenspace characteristics at each instant of time by providing solution to a class of generalized EVP, as opposed to joint diagonalization performed through SOBI. It can therefore, be anticipated that the recursive formulation of CCA will be computationally less intensive. To obtain recursive eigenspace estimates, the individual lower dimensional matrices shown in Eq. 7.1 needs to be updated at each time stamp. In this context, the response covariance matrix $\mathbf{C}_{uu}(k)$ at any instant k can be expressed as a function of the covariance matrix at the previous time stamp, $\mathbf{C}_{uu}(k-1)$ and the response vector U_k at the kth instant, according to:

$$\mathbf{C}_{uu}(k) = \frac{k-1}{k}\mathbf{C}_{uu}(k-1) + \frac{1}{k}U_k U_k^T \tag{7.9}$$

As previously mentioned, the FOEP based approach is a versatile idealization that does not take into consideration the nature of the dataset. The applicability of the method for nonstationary datasets involves the consideration of the recursive mean at each instant of time. The covariance estimate for cases involving recursive mean update ($\mu_k = \frac{k-1}{k}\mu_{k-1} + \frac{1}{k}U_k$) can be expressed as:

$$\tilde{\mathbf{C}}_{uu}(k) = \frac{k-1}{k}\mathbf{\Sigma}_k^{-1}\mathbf{\Sigma}_{k-1}\tilde{\mathbf{C}}_{uu}(k-1)\mathbf{\Sigma}_{k-1}\mathbf{\Sigma}_k^{-1} + \mathbf{\Sigma}_k^{-1}\Delta\mu_k\Delta\mu_k^T\mathbf{\Sigma}_k^{-1} + \frac{1}{k}\left[\mathbf{U}_k - \mu_k\right]\left[\mathbf{U}_k - \mu_k\right]^T \tag{7.10}$$

The eigen decomposition of the covariance estimate obtained in Eq. 7.10 can be expressed as:

$$\tilde{\mathbf{W}}_k\tilde{\Upsilon}_k\tilde{\mathbf{W}}_k^T = \frac{k-1}{k}\mathbf{\Sigma}_k^{-1}\mathbf{\Sigma}_{k-1}\tilde{\mathbf{C}}_{uu}(k-1)\mathbf{\Sigma}_{k-1}\mathbf{\Sigma}_k^{-1} + \mathbf{\Sigma}_k^{-1}\Delta\mu_k\Delta\mu_k^T\mathbf{\Sigma}_k^{-1} + \frac{1}{k}\left[\mathbf{U}_k - \mu_k\right]\left[\mathbf{U}_k - \mu_k\right]^T \tag{7.11}$$

The POC vector at the kth time instant can be estimated as: $\tilde{\psi}_k = \tilde{\mathbf{W}}_{k-1}^T U_k$. Carrying out proper substitutions in Eq. 7.11 and scaling the data to unit variance, one obtains:

$$\tilde{\mathbf{W}}_k\tilde{\Upsilon}_k\tilde{\mathbf{W}}_k^T = \frac{k-1}{k}\tilde{\mathbf{W}}_{k-1}\tilde{\Upsilon}_{k-1}\tilde{\mathbf{W}}_{k-1}^T + \Delta\mu_k\Delta\mu_k^T + \frac{1}{k}\left[\mathbf{W}_{k-1}\tilde{\psi}_k - \mu_k\right]\left[\mathbf{W}_{k-1}\tilde{\psi}_k - \mu_k\right]^T \tag{7.12}$$

For structural systems evolving from zero mean excitation, the above equation can be simplified as:

$$\tilde{\mathbf{W}}_k k\tilde{\Upsilon}_k\tilde{\mathbf{W}}_k^T = \tilde{\mathbf{W}}_{k-1}\left[(k-1)\tilde{\Upsilon}_{k-1} + \tilde{\psi}_k\tilde{\psi}_k^T\right]\tilde{\mathbf{W}}_{k-1}^T \tag{7.13}$$

For the RCCA algorithm to be stable and robust, it is important that the term $[(k-1)\tilde{\Upsilon}_{k-1} + \tilde{\psi}_k\tilde{\psi}_k^T]$ should have a diagonally dominant structure. This ensures the use of Gershgorin's theorem to evaluate the EVD of this term for low to moderate values of damping [14]. Again, $\tilde{\psi}_k\tilde{\psi}_k^T$ represents the correlation between the POC estimates at a particular instant. The scaled covariance between POC estimates can be expressed as follows:

$$\tilde{\psi}_k\tilde{\psi}_k^T = \tilde{\mathbf{W}}_{k-1}^T u_k {u_k}^T \tilde{\mathbf{W}}_{k-1} = q_{k-1}{q_{k-1}}^T + \gamma {q_{k-1}}^T + \gamma^T q_{k-1} + \gamma\gamma^T \tag{7.14}$$

It can be inferred that the term $q_{k-1}{q_{k-1}}^T$ represents a diagonally dominant structured matrix, which in turn ensures the diagonal dominance of the term

$\tilde{\psi}_k\tilde{\psi}_k^T$. Reverting to Eq. 7.13, it can therefore, be concluded that the matrix $[(k-1)\tilde{\Upsilon}_{k-1}+\tilde{\psi}_k\tilde{\psi}_k^T]$ becomes diagonally dominant which ensures a straightforward application of Gershgorin's theorem, leading to the recursive eigenspace updates through RCCA:

$$\left.\begin{aligned}\tilde{\mathbf{W}}_k &= \tilde{\mathbf{W}}_{k-1}\mathbf{H}_k\\ \tilde{\Upsilon}_k &= \frac{\tilde{\Lambda}_k}{k}\end{aligned}\right\} \tag{7.15}$$

The recursive algorithm of Eq. 7.9 is transformed to obtain the values of H_k and Λ_k. As the term $(k-1)\tilde{\Upsilon}_{k-1}+\tilde{\psi}_k\tilde{\psi}_k^T$ is diagonally dominant, the eigen values can be considered to be the diagonal entries of the matrix. Therefore, the ith diagonal entry of the term Λ_k can be represented as $(k-1)\lambda_i+\tilde{\psi}_i^2$, with λ_i as the (i,i) element of $\tilde{\Upsilon}_{k-1}$ and $\tilde{\psi}_i$ as the ith entry of the POC estimate. Once the eigen vectors corresponding to each eigen value is obtained, H_k can be subsequently evaluated. This discussion provides an insight towards obtaining recursive eigenspace updates for structural responses oriented in an uniaxial direction. However, it should be noted that the block covariance matrix shown in Eq. 7.1 consists of the covariance matrices obtained from the responses for multi-directional dependencies as well. Therefore, the structure of the fundamental matrices involved in the EVD of Eq. 7.3 can be expressed as:

$$\mathbf{A}(k)=\begin{bmatrix}\mathbf{0} & \mathbf{C}_{uv}(k)\\ \mathbf{C}_{vu}(k) & \mathbf{0}\end{bmatrix},\quad \mathbf{B}(k)=\begin{bmatrix}\mathbf{C}_{uu}(k) & \mathbf{0}\\ \mathbf{0} & \mathbf{C}_{vv}(k)\end{bmatrix} \tag{7.16}$$

The matrices **A** and **B** can be recursively updated through the recursive updates of their individual covariance estimates by following the Eq. 7.9. The structure for the recursive update of the matrices **A** and **B** can be written as:

$$\left.\begin{aligned}\mathbf{A}(k) &= \frac{k-1}{k}\begin{bmatrix}\mathbf{0} & \mathbf{C}_{uv}(k-1)\\ \mathbf{C}_{vu}(k-1) & \mathbf{0}\end{bmatrix}+\frac{1}{k}\begin{bmatrix}\mathbf{0} & U_kV_k^T\\ V_kU_k^T & \mathbf{0}\end{bmatrix}\\ \mathbf{B}(k) &= \frac{k-1}{k}\begin{bmatrix}\mathbf{C}_{uu}(k-1) & \mathbf{0}\\ \mathbf{0} & \mathbf{C}_{vv}(k-1)\end{bmatrix}+\frac{1}{k}\begin{bmatrix}U_kU_k^T & \mathbf{0}\\ \mathbf{0} & V_kV_k^T\end{bmatrix}\end{aligned}\right\} \tag{7.17}$$

The above mentioned equation set can be rewritten in terms of recursive estimates that are updated at each instant of time. Recasting Eq. 7.17 in a recursive format, one obtains:

$$\begin{aligned}\mathbf{A}(k) &= \frac{k-1}{k}\mathbf{A}(k-1)+\frac{1}{k}\mathcal{U}^{(U,V)}(k)\\ \mathbf{B}(k) &= \frac{k-1}{k}\mathbf{B}(k-1)+\frac{1}{k}\mathcal{U}^{(U,U)}(k)\end{aligned} \tag{7.18}$$

Where,

$$\mathcal{U}^{(U,V)}=\begin{bmatrix}\mathbf{0} & U_kV_k^T\\ V_kU_k^T & \mathbf{0}\end{bmatrix}\quad and\quad \mathcal{U}^{(U,U)}=\begin{bmatrix}U_kU_k^T & \mathbf{0}\\ \mathbf{0} & V_kV_k^T\end{bmatrix}$$

One of the key features of the FOEP based approaches lies in its significant reduction of memory consumption and time complexities. It can be well understood that performing an iterative estimation in the backdrop of a generalized eigenvalue problem is more involved and complicated. Additionally, a thorough survey of the literature reveals that implementing step wise updates of the covariance matrix at each iteration is cumbersome [13–15]. This necessitates that efficient numerical techniques should be resorted to for such situations. Based on this premise, consider a simplified version of Eq. 7.3 at the kth time instant, given by:

$$\mathbf{A}(k)\hat{\mathbf{w}}(k) = \rho(k)\mathbf{B}(k)\hat{\mathbf{w}}(k) \tag{7.19}$$

At this stage, it can very well be presumed that Eq. 7.19 needs to be recursively updated for every sample. As evident from Eq. 7.15, the rank-one update of the covariance matrix provides the recursive estimations at each instant. Using a similar approach, consider the rank-one perturbation of the eigen decomposition given by Eq. 7.19:

$$(\mathbf{A}+\Delta\mathbf{A})\,(\hat{\mathbf{w}}+\Delta\hat{\mathbf{w}}) = (\rho+\Delta\rho)\,(\mathbf{B}+\Delta\mathbf{B})\,(\hat{\mathbf{w}}+\Delta\hat{\mathbf{w}}) \tag{7.20}$$

The key step in recursively updating the above equation can be inferred from the development of Eq. 7.17 using key covariance matrices. This clearly demonstrates that the eigenspace updates can be eventually obtained through the FOEP technique without actually performing EVD on the subsequent data points, thereby reducing the time complexity and memory consumption, to an great extent. On expanding term by term, Eq. 7.20 becomes:

$$\begin{aligned}
&\begin{bmatrix} \mathbf{0} & \frac{k}{k+1}\mathbf{C}_{uv}(k)+\frac{1}{k+1}U_{k+1}V_{k+1}^T \\ \frac{k}{k+1}\mathbf{C}_{vu}(k)+\frac{1}{k+1}V_{k+1}U_{k+1}^T & \mathbf{0} \end{bmatrix}(k+1)\,[\mathbf{W}_k\mathbf{H}_{k+1}] \\
&= \Lambda_{k+1}\begin{bmatrix} \frac{k}{k+1}\mathbf{C}_{uu}(k)+\frac{1}{k+1}U_{k+1}U_{k+1}^T & \mathbf{0} \\ \mathbf{0} & \frac{k}{k+1}\mathbf{C}_{vv}(k)+\frac{1}{k+1}V_{k+1}V_{k+1}^T \end{bmatrix}[\mathbf{W}_k\mathbf{H}_{k+1}] \\
&= \frac{\Lambda_{k+1}}{k+1}\left\{\begin{bmatrix} k\mathbf{C}_{uu}(k) & \mathbf{0} \\ \mathbf{0} & k\mathbf{C}_{vv}(k) \end{bmatrix}+\begin{bmatrix} U_{k+1}U_{k+1}^T & \mathbf{0} \\ \mathbf{0} & V_{k+1}V_{k+1}^T \end{bmatrix}\right\}[\mathbf{W}_k\mathbf{H}_{k+1}]
\end{aligned} \tag{7.21}$$

It can be understood that in order to obtain the eigenspace at each sample point, Eq. 7.20 needs to be solved, which requires a complete solution of a generalized eigen value problem. In this regard, the authors have resorted to the use of Woodbury's inverse identity that can provide the inverse of any full rank positive definite matrix. The inverse EVD at the $k+1^{th}$ step, therefore, becomes:

$$\hat{\mathbf{w}}(k+1) = \rho(k+1)\mathbf{A}^{-1}(k+1)\mathbf{B}(k+1)\hat{\mathbf{w}}(k+1) \tag{7.22}$$

Using Woodbury's identity, the inverse of the matrix can be computed as follows:

$$\begin{aligned}
\mathbf{A}(k+1) &= \begin{bmatrix} \mathbf{0} & \frac{k}{k+1}\mathbf{C}_{xy}(k)+\frac{1}{k+1}X_{k+1}Y_{k+1}^T \\ \frac{k}{k+1}\mathbf{C}_{yx}(k)+\frac{1}{k+1}Y_{k+1}X_{k+1}^T & \mathbf{0} \end{bmatrix} \\
&= \frac{k}{k+1}\mathbf{A}(k)+\frac{1}{k+1}\begin{bmatrix} \mathbf{0} & X_{k+1}Y_{k+1}^T \\ Y_{k+1}X_{k+1}^T & \mathbf{0} \end{bmatrix} \\
&= \mathbf{P}+\mathbf{QR}
\end{aligned} \tag{7.23}$$

The matrix product **QR** can be expressed as:

$$\begin{aligned}\mathbf{QR} &= \frac{1}{k+1}\begin{bmatrix} \mathbf{0} & U_{k+1}V_{k+1}^T \\ V_{k+1}U_{k+1}^T & \mathbf{0} \end{bmatrix} \\ &= \frac{1}{k+1}\begin{bmatrix} \mathbf{0} & U_{k+1} \\ V_{k+1} & \mathbf{0} \end{bmatrix}\begin{bmatrix} U_{k+1}^T & \mathbf{0} \\ \mathbf{0} & V_{k+1}^T \end{bmatrix} \\ &= \begin{bmatrix} \mathbf{0} & \frac{U_{k+1}}{k+1} \\ \frac{V_{k+1}}{k+1} & \mathbf{0} \end{bmatrix}\begin{bmatrix} U_{k+1}^T & \mathbf{0} \\ \mathbf{0} & V_{k+1}^T \end{bmatrix}\end{aligned} \tag{7.24}$$

On obtaining these individual matrices and their simpler variations, the Woodbury's identity can now be implemented to obtain the inverse of the matrix **A**, in real time, as follows:

$$\begin{aligned}\mathbf{A}^{-1}(k+1) &= [\mathbf{P}(k+1)+\mathbf{Q}(k+1)\mathbf{R}(k+1)]^{-1} \\ &= \mathbf{P}^{-1}(k+1) - \mathbf{P}^{-1}(k+1)\mathbf{Q}(k+1) \\ &\quad \left[\mathbf{R}^{-1}(k+1)+\mathbf{P}^{-1}(k+1)\mathbf{Q}(k+1)\right]^{-1}\mathbf{P}^{-1}(k+1)\end{aligned} \tag{7.25}$$

On carrying out proper substitutions, the recursive EVD can be solved, in order to obtain the eigenspace updates in real time. These updates are further utilized to estimate the LNMs of the system.

7.4 Proposed algorithm

The focus of this chapter is to outline RCCA as an effective online modal identification approach. Addition to this is the transformation of the algorithm to a damage detection algorithm, which has also been addressed in detail in the previous chapter. The key difference between the approaches adopted lie in the fact that RCCA based recursive modal ID automatically employs its online damage detection tool; in contrast to the stand-alone approach (of RCCA as an online damage monitoring tool) adopted in the previous chapter.

For RCCA to function as a real time modal identification approach, the estimated LNMs are compared against the theoretical modes to investigate their correlation and subsequently, the transformed responses are obtained on which the TVAR models are fit, for detecting damage. On ascertaining the instant of damage, the spatial module is invoked where the spatial RRE is tracked over each DOF to identify the location of damage. A distortion in the RRE plots confirms the exact location of damage in the structure. For an easy comprehension of the detailed process, the basic steps of the algorithm are enumerated as follows:

1. First, a block covariance matrix incorporating the individual auto- and cross-covariance matrices is fabricated from the set of physical responses. Traditional CCA is applied on some initial data points (around 100 in number) in order to estimate the initial eigenspace. The number of data

points chosen here is arbitrary and considered only to stabilize the algorithm for subsequent real time damage detection. The batch implementation comprises the EVD updates shown in Eq. 7.2, yielding the eigenspace represented by ρ and $\hat{\mathbf{w}}$.

2. The recursive gain depth parameter is employed to estimate the covariance estimate at the present time instant using the covariance estimate at preceding time instant. From these recursive updates, the eigenvector and eigenvalue matrices are updated using FOEP approach and the transformed responses (principal components) are obtained using the RCCA algorithm. While the covariance updates for the matrices are given by Eq. 7.17, the iterative set of eigenspace updates carried out after the initial batch implementation can be easily obtained through Eq. 7.15.

3. The eigenspace updates so obtained are investigated for cases for permutation ambiguity in real time. On sorting the eigenvectors corresponding to each of the eigenvalues, the LNMs are estimated and compared against the theoretical modes. The modal assurance criterion (MAC) is used to establish the correlation between the estimated and theoretical modes, and is subsequently reported.

The aforementioned steps for the proposed method are shown in the form of a stepwise flowchart in Fig. 7.1. The recursive MAC values evolving over time

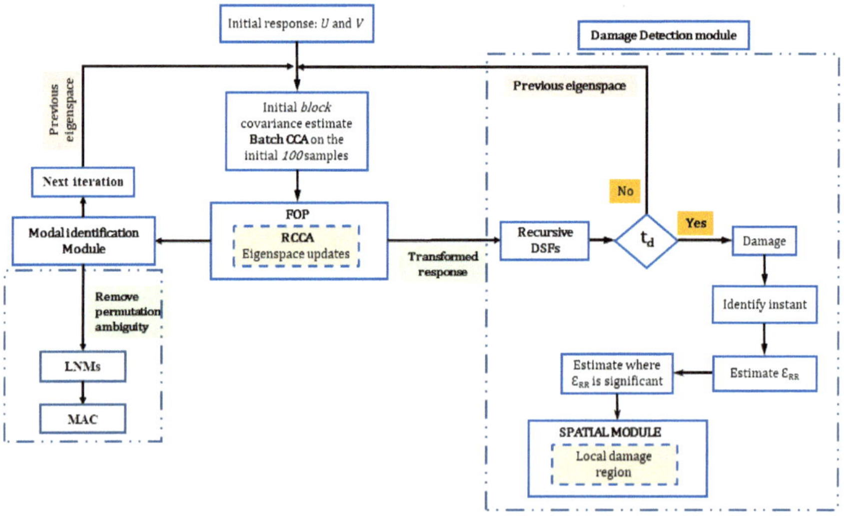

Figure 7.1: Workflow diagram of modal identification and damage detection module of proposed technique.

indicate that the proposed method is well suited to extract the modes in real time. A few important characteristics of the proposed algorithm include: (i) implementation on the streaming dataset; signifying that the algorithm essentially operates in *real time* (ii) detection of damage instant and its location without ad-hoc windowing or gathering of any sort of baseline data, making it *baseline-free* and (iii) absence of parameters controlling the functioning of the algorithm, rendering it *parameter independent*. These characteristics of the proposed algorithm make it an ideal candidate for real time structural modal identification and damage detection framework.

7.5 Model description

To demonstrate the efficacy of the proposed algorithm towards simultaneous modal identification and damage detection, three numerical models are considered in this paper: (a) *a 5–DOF linear system*, (b) *a 5–DOF system with a non-linear Bouc-Wen (B-W) isolator lumped at the base*. Numerical simulations carried out on these models exhibit the wide applicability of the proposed RCCA technique for identification of modal parameters for linear systems and also simultaneously detect damage for strongly non-linear systems. The system properties, nature and duration of excitations and the identification results are discussed in the following subsections.

7.5.1 5 DOF mass, spring and dashpot system: Model description

In order to check the robustness of the proposed algorithm numerical simulations are carried out on a 5–DOF mass, spring and dashpot system. The equation of motion of the vibrating system and its state-space formulation can be found in [44]. Following the subsequent steps of mode extraction as previously discussed, the structural modal response can be obtained.

$$\mathbf{M}\ddot{X}(t) + \mathbf{C}\dot{X}(t) + \mathbf{K}X(t) = \mathbf{F}(t) \tag{7.26}$$

where, $X(t)$ is a displacement vector at the degrees of freedom, **M**, **C** and **K** are symmetric mass, damping and stiffness matrix, respectively. $\mathbf{F}(t)$ is the excitation force vector. The solution of the Eq. 7.26 can be written in terms of expansion of vibration modes. The state-space representation for the system subjected to a force vector $F(t)$ can be written as,

$$\begin{aligned} \dot{U} &= \mathbf{A}U + \mathbf{B}F \\ Y &= \mathbf{C}U \end{aligned} \tag{7.27}$$

where, U = vector of states and Y = system response vector governed by the **C** matrix. The system matrix, **A**, and the excitation matrix **B** are given by,

$$\mathbf{A} = \begin{bmatrix} [\mathbf{0}]_{5\times5} & [\mathbf{I}]_{5\times5} \\ -\mathbf{M}^{-1}\mathbf{K} & -\mathbf{M}^{-1}\mathbf{C} \end{bmatrix}$$
$$\mathbf{B} = \begin{bmatrix} 0 & 0 & 0 & 0 & 0 & -\frac{1}{m} & -\frac{1}{m} & -\frac{1}{m} & -\frac{1}{m} & -\frac{1}{m} \end{bmatrix}^T \tag{7.28}$$

The mass of the system is considered as $10kg$ at each floor level and the stiffness of the individual floors are assumed to be $2kN/m$. The damping ratio for the system is kept at $\zeta = 2.0\%$ critical for each mode. The system is simulated using zero mean Gaussian white noise excitation applied to each floor, excited for a duration of $100s$ at a sampling frequency of $100Hz$. The MAC values are employed to check the correlation between the LNMs and theoretical modes, mathematically defined in Eq. 7.29. The MAC values provide an estimate of the correlation between the theoretical modes and LNMs obtained through recursive modal identification using RCCA. The estimated mode shapes are expected to converge to the theoretical modes with the progression of samples, in real time. The expression to obtain MAC values are shown as follows:

$$MAC_i = \frac{(\phi_i^T \varphi_i)^2}{(\phi_i^T \phi_i)(\varphi_i^T \varphi_i)} \tag{7.29}$$

Case studies are considered where the modes are recursively identified and the damage induced to the system is recursively determined, presented in a single framework in the later stages of the paper.

7.5.2 5 DOF B-W model description

A numerically simulated 5 storied structure with 4 floors of linear stiffness and a nonlinear B-W isolator lumped at the base is considered for damage detection studies. The model under consideration has been adopted from Krishnan et al. [14]. A lead rubber bearing isolator (LRB) separates the base from the surrounding ground. The governing differential equation of motion of the system can be represented as:

$$\mathbf{M}\ddot{U} + \mathbf{C}\dot{U} + \mathbf{K}U - \kappa z Q_{pb} = k_b U_b + c_b \dot{U}_b - \mathbf{MI}\ddot{\mathbf{U}}_\mathbf{g} \tag{7.30}$$

where $Q_{pb} = \left(1 - \frac{k_{yield}}{k_{initial}}\right) Q_y$ and k_b and c_b are the stiffness and the viscous damping respectively, in the horizontal direction. The numerical values of the matrix elements can be found in [14]. In Eq. (7.30), the vector I consists of all elements as unity and $\ddot{\mathbf{U}}_\mathbf{g}$ represents the ground acceleration. The vector U represents the displacement of each floor and the base. It should be noted that the forces due to base damping and stiffness terms (k_b and c_b) are included in the nonlinear force (F) due to the LRB base isolator. The term $k_{initial}$ represents the initial shear stiffness and k_{yield} is the post yield shear stiffness of the LRB. The evolutionary variable z is used to provide the hysteretic component of the horizontal force, $Q_{hyst} = zQ_{pb}$. The variable z can be obtained by solving the following nonlinear differential equation:

$$\dot{z} = -\gamma z \left|\dot{U}_b\right| \left|z^{n-1}\right| - \beta \dot{U}_b \left|z^n\right| + A\dot{U}_b \tag{7.31}$$

Here γ, β, A and n are the shape parameters for the hysteresis loop [13, 14]. The nonlinear parameter κ introduced in Eq. (7.30) controls the amount of nonlinearity in the system. The alteration in the nonlinear parameter corresponds to a global change in the system. For instance, a change in κ from 1 to 0.4 can be interpreted to be a 60% change in the nonlinear force term of the system. This force induction through a change in the value of κ corresponds to a damage in the present context, as it contributes to a change in the nonlinearity of the system. An important aspect of the damage induction through changes in nonlinear force term is that the changes are permanent and do not wear over time, and thus, it can be safely assumed that the damage is certainly not repaired as time progresses. To carry out a real time damage detection study, the system is simulated using zero mean Gaussian white noise excitation applied to each floor, excited for a duration of $100s$ at a sampling frequency of $100Hz$. The nonlinear equations are time integrated using Ito-Taylor numerical schemes. The readers are referred to [44, 45] for details.

7.6 Modal identification results

In this module, the modal identification results obtained from the numerical models discussed in the Section 7.5 is presented. The results are studied in order to understand the efficacy of the proposed technique in capturing the modal parameters in real time. The acceleration responses of the systems are considered for simultaneous modal identification and subsequent damage detection. The batch identification results using traditional CCA for 5 DOF system is presented, followed by the recursive estimation for the models. Additionally, the performance of SOBI as an established method in the genre of modal identification is also assessed.

7.6.1 Modal identification results using batch CCA

The objective of this section is to check the performance of the proposed RCCA technique against its batch counterpart and to assess the attributes of the LNMs through numerical simulations carried out on the 5–DOF system. The complete set of numerically simulated responses is taken into account for performing batch analysis through CCA, considered in growing windows. Accumulative windows of 2000 samples each are taken into account for analysis through batch CCA and the LNMs corresponding to each window is obtained. The correlation between the estimated and the theoretical modes are evaluated using MAC values and

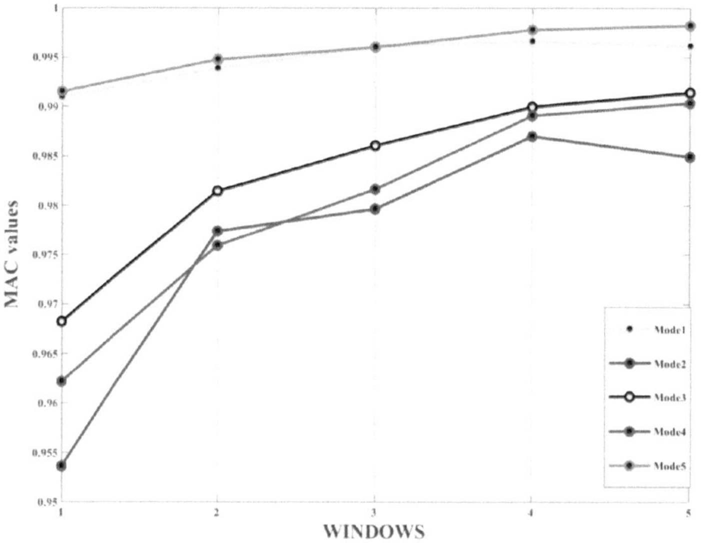

Figure 7.2: Evolution of MAC values for 5–DOF system using batch CCA.

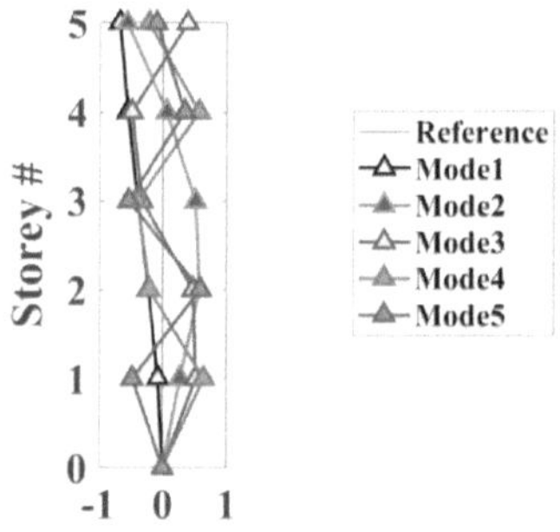

Figure 7.3: Estimated LNMs at each window for 5–DOF system using batch CCA.

presented in the Fig. 7.2. It can be observed from Fig. 7.3 that the MAC values attain almost a value close to unity, which implies a near perfect correlation of the LNMs with the theoretical modes. The plots of the LNMs for each window shown in Fig. 7.3 illustrates that the estimated modes converge to the true modes with the accumulation of sample points using batch CCA technique.

Figure 7.3 portrays the correlation of the estimated LNMs and theoretical mode for each window. The theoretical mode is indicated by a thick dotted red line and the LNM corresponding to 5th window is indicated by a thick green line with star marker. From the figure it can be visually inferred that the estimated LNMs obtained from the 5th window is found to correlate with the theoretical mode perfectly. The convergence of the estimated LNMs to the theoretical modes in incremental windows, along with the progression of the samples, demonstrates that the CCA algorithm is a good candidate for modal identification of vibrating systems. However, it should be noted that the batch implementation of CCA is associated with some key shortcomings as (i) continuous vibration based monitoring is not possible due to the requirement of batches of data at regular intervals, that leads to (ii) increase in computational exhaustion due to the necessity of incremental windows for subsequent batch analysis. These drawbacks motivate the utilization of the proposed RCCA algorithm that allows continuous monitoring of structures with less computational load.

7.6.2 Modal identification results using RCCA

Previous identification of structural modal parameters using batch CCA demonstrated the utility of the method in extracting the approximated vibrating modes of the system. Due to the aforementioned shortcomings of the batch formulation, it becomes necessary to investigate the problem of modal identification in the light of a recursive implementation that can provide estimations at each instant. This greatly motivates the development of RCCA towards identifying the modal parameters, which holds a major entitlement of the work. The implementation of the proposed RCCA technique ensures that the estimation of the vibrating modes occur at each instant, that forms the sole basis of continuous monitoring, as opposed to traditional batch schemes that resort to batches of data for further post-processing. The estimation of LNMs in real time and demonstrating its near perfect correlation with the theoretical modes provides a formidable challenge that has not been attempted in the literature so far. The ensuing section provides the identification results for the previously mentioned linear 5 DOF system, in order to ensure the effectiveness of the proposed technique and maintain parity throughout.

7.6.2.1 Recursive modal identification results for 5–DOF linear system

In this section, the applicability of the proposed RCCA method is examined using the aforementioned 5 DOF linear system. The acceleration responses obtained from the system is taken into consideration on which batch CCA for the first 100 samples are applied, required to stabilize the algorithm. The block covariance matrix generated from the physical responses is used for subsequent recursive estimation of the eigenspace at each instant of time. The estimated LNMs obtained using RCCA approach is checked for recursive correlation with the theoretically obtained modes, through MAC values examined in a recursive fashion. The recursive evolution of the MAC values at each sample point is shown in Fig. 7.4.

It can be clearly observed from the plots that the recursively estimated MAC values provide near perfect correlation with the theoretical mode, corresponding to a converged MAC plot. The commencement of the convergence of the MAC values to near unity can be approximately observed at *18%* of the total sample size, which is an important realistic finding, considering the recursive nature of the implementation. Previous attempts on estimation of modes through EASI [12] has provided an initial convergence at *30%* sample population using a semi-automated technique that involves considerable batch processing of the dataset. Additionally, to establish convergence in attaining near-perfect correlation with the progression of streaming samples, a separate plot illustrating the recursive estimation at pre-selected sample points is provided in Fig. 7.5. The plot clearly shows that the initial estimates of the modes are somewhat erroneous, owing to the recursive nature of the RCCA implementation, but converge to the actual mode as more samples are streamed online, which stabilizes the algorithm. As observed from the figure, the LNM obtained at the 10000th sample point accurately converges with the theoretical mode that not only establishes the near-perfect convergence of the modes but also justifies the use of RCCA as an efficient modal identification technique. As the presence of *sensor noise* is inevitable, estimation of the LNMs corresponding to a 10% noise level is evaluated. Recursive MAC values corresponding to pre-selected Monte Carlo iterations is provided in Table 7.2. The noise level considered is adequate, as the focus of recursive modal identification is centred towards numerically simulated discrete dynamical systems. From the table, it can be inferred that recursive estimation of LNMs for noisy sensors contribute to approximately 1% error (in comparison to a noise-free environment), which is acceptable, from an engineering perspective and real time identification standpoint.

Based on the key findings through these discussions, it is now clear that the RCCA algorithm is well equipped to identify the modal parameters of a vibrating system.

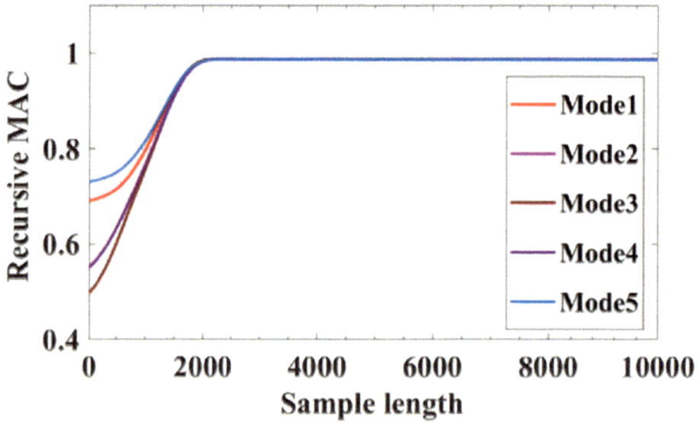

Figure 7.4: Recursive evolution of MAC across samples for 5–DOF system.

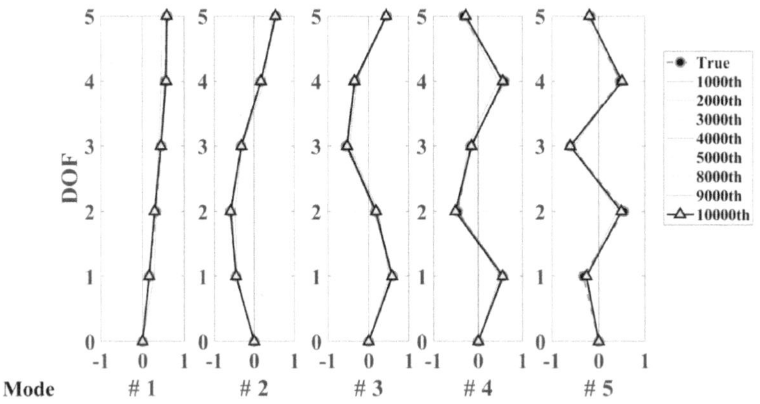

Figure 7.5: Comparison between the theoretical mode and LNM for 5–DOF system using RCCA.

Table 7.2: Iteration vs MAC value for 5–DOF linear system for 10% sensor noise.

i^{th} iter	MAC_1	MAC_2	MAC_3	MAC_4	MAC_5
1000	0.785	0.758	0.763	0.766	0.805
2000	0.966	0.968	0.968	0.969	0.968
3000	0.969	0.967	0.962	0.967	0.967
4000	0.967	0.967	0.965	0.967	0.967
5000	0.968	0.973	0.966	0.966	0.969
8000	0.969	0.969	0.965	0.971	0.967
9000	0.969	0.968	0.966	0.967	0.966
10000	0.968	0.969	0.970	0.966	0.965

The identification of modal parameters is extended for different values of modal damping in order to check the robustness of the algorithm for moderate to higher damping environments. The MAC values presented in Table 7.3 are the mean of the MAC values obtained when the MAC of each mode at i^{th} iteration (approximately at 18% of the entire sample population) is greater than 0.95 (shown in the Fig. 7.4). Tablc 7.3 compares the modal identification results of the proposed RCCA algorithm for damping ratios ranging from 1% to 5%. According to Table 7.3, the identification of modal parameters is observed to be accurate for all the ζ, which is clearly advantageous over the ICA based approaches [16,23]. Moreover, the estimation of near-prefect natural frequencies over the ensemble average for varying damping levels suggests the accuracy of the proposed RCCA approach. Therefore, from the results in Table 7.3 it can be inferred that the identification of modal parameters is both robust and accurate (i.e., MAC values are close to unity) for moderately damped systems ($\zeta \leq 5\%$).

Table 7.3: Accuracy (MAC) of the identification using proposed RCCA algorithm for 5–DOF system.

Vibration Modes	**Input Damping Ratios (%) and Natural Frequencies (Hz)**									
	ζ = **1.0**	ω_n	ζ = **2.0**	ω_n	ζ = **3.0**	ω_n	ζ = **4.0**	ω_n	ζ = **5.0**	ω_n
Mode 1 (MAC_1)	0.9878	0.65	0.9877	0.64	0.9878	0.64	0.9874	0.64	0.9875	0.64
Mode 2 (MAC_2)	0.9775	1.86	0.9778	1.87	0.9778	1.87	0.9776	1.86	0.9775	1.85
Mode 3 (MAC_3)	0.9769	2.95	0.9763	2.93	0.9661	2.90	0.9766	2.93	0.9666	2.92
Mode 4 (MAC_4)	0.9759	3.75	0.9757	3.74	0.9756	3.74	0.9648	3.72	0.9648	3.72
Mode 5 (MAC_5)	0.9868	4.31	0.9867	4.31	0.9867	4.31	0.9865	4.30	0.9862	4.29

Based on the above discussions, it can be clearly understood that the proposed RCCA algorithm provides eigenspace updates at each instant of time. The estimation of structural damping becomes difficult when attempted in a real time framework. To maintain parity, damping estimation is attempted in a *semi-automated* approach using the widely popular Natural Excitation Technique (NExT) [46]. windows of data corresponding to pre-selected sample points are considered from the streaming data which provides damping estimates in a near-recursive framework. In this context, *overlapping* windows corresponding to 2000 sample points each are considered and the estimation results are presented in Table 7.4. This implies that the first window consists of 2000 samples, the second of 4000 samples and the final window considers the entire ensemble population. From a theoretical standpoint, the performance of the proposed algorithm in the presence of higher damping can be used as a yardstick of its efficacy and therefore, input damping ratios varying from 2-5% are considered. It can be inferred from the table that the estimated damping ratios accurately correspond to the input damping values with the progression of the samples from the streaming dataset. The robustness of the algorithm to accurately estimate damping can be verified from this analysis. In order to demonstrate the applicability of the method towards real time damage detection, the eigenspace updates are now provided as input to the damage detection module, as illustrated in Fig. 7.1. The transformed response obtained at each time stamp is fit using TVAR models of appropriate model order. Further, these responses are processed by a set of recursive DSFs that indicate the exact instant of damage to the system. Once the damage is detected, the focus now shifts onto the spatial detection module where the location of damage is precisely identified in a simultaneous recursive framework. The detailed working steps are illustrated in Fig. 7.1 for an easy comprehension of the above discussion.

7.7 Real time damage detection: An application of RCCA

Damage can be defined as the alteration in structural physical parameters such as mass, stiffness and damping, that can also be inferred from the changes in mode shapes, modal curvatures and modal responses as well [19, 33, 38–42]. The primary objective of adopting the same numerically simulated systems is to appreciate the unification of recursive modal identification and damage detection in a single framework using RCCA, which forms the key entitlement of the current work. Additionally, to demonstrate the effectiveness of the proposed algorithm towards detecting damage in real time for nonlinear systems, the 5 DOF B-W system is considered. The key DSFs used in this work to infer damage is first discussed followed by The detailed discussion on the real time detection results using the proposed RCCA algorithm and its batch counterpart.

Table 7.4: Windowed estimates of damping–A semi-automated approach.

Window	Input damping	Estimated damping	Input damping	Estimated damping	Input damping	Estimated damping
		Mean (coeff. of var. (%))		**Mean** (coeff. of var. (%))		**Mean** (coeff. of var. (%))
#1	0.02	**0.0178** 1.005	0.03	**0.0293** 0.916	0.05	**0.0494** 0.905
#2	0.02	**0.0194** 0.564	0.03	**0.0295** 0.557	0.05	**0.0497** 0.551
#3	0.02	**0.0197** 0.454	0.03	**0.0298** 0.045	0.05	**0.0498** 0.045
#4	0.02	**0.02** 0.043	0.03	**0.0298** 0.043	0.05	**0.0497** 0.043
#5	0.02	**0.02** 0.004	0.03	**0.03** 0.004	0.05	**0.05** 0.004

7.7.1 Recursive damage sensitive features

The proposed RCCA based scheme updates the eigenspace at each instant of time. This eigenspace can be further utilized to identify the damage induced to the system due to forced excitations. However, the eigenspace by itself is inadequate in exhibiting the damage patterns of the system. Therefore, certain damage markers, known as DSFs, are employed to ascertain the presence of damage and its location. The key characteristics of a good DSF lies in its potential to detect the presence of damage, locate and estimate the severity of damage and effectively distinguish between the damaged and undamaged states of the structure. Further, the additional attributes of these DSFs include their ability to function *online*, in order to identify the change of states in real time [13–15]. In this context, the aforementioned TVAR coefficients and the RRE vectors are utilized as appropriate DSFs for identifying the spatio-temporal patterns of damage in time. The detailed derivation for the same has already been encountered in *chapter 3* and hence, not reported here for brevity.

7.7.2 Damage detection results for linear 5–DOF system

The applicability of the proposed method towards identifying damage for a linear system can be examined by providing the streaming acceleration response as input to the algorithm. The aim of the present section is to comparatively assess the performance of the proposed RCCA method against a well-established identification technique, SOBI [16]. As the comparison of the proposed recursive scheme against the established offline method is not justified, it becomes necessary to introduce windows to analyze data in batches for the SOBI method. Incremental windows of 1000 samples of the physical response are considered on to which an analogous residual error (ARE) is evaluated [45]. It should be noted that the choice of windows is purely arbitrary and does not play any role in the functioning of the algorithm. In this regard, the damage to the linear 5 DOF system is considered as a 25% storey stiffness change at 31s from the start of the excitation. The linear storey stiffness of the third floor for the system is considered to undergo a change that accounts for a local damage to the entire structure. As the damage is confined only to a single storey of the system, it is termed as '*local*' damage, in the present context. Damage detection through RCCA is illustrated by the use of local RRE, which is an effective indicator of identifying spatial patterns of damage [13]. The windowed SOBI technique analyzes the data in batches that yields ARE plot corresponding to each window. On the contrary, the RCCA method investigates each sample of data and generates the corresponding eigenspace, in real time. The detection results using both the methods are compared in Fig. 7.6.

It can be observed from the figure that the ARE plot shows a consistent increase in the mean levels for the first two windows. On reaching the third window, the mean shift is inconsistent with the previous windows for the time

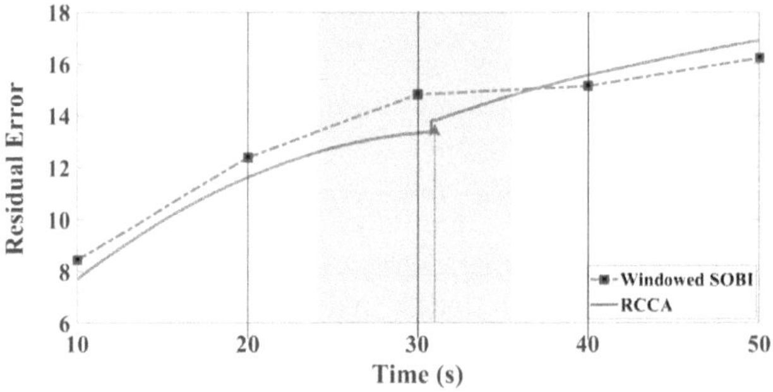

Figure 7.6: Damage detection for 5–DOF system for 25% change in linear stiffness at 3rd floor.

range of 30-40s, that corresponds to a possible occurrence of an event. Contrary to the ARE plot, the RRE examined at each instant of time, shows distortion at 31s from the start, thereby confirming the occurrence of damage at that instant. This verifies that the proposed method is adept at identifying spatial damage for a linear system. While the batch formulation using windowed SOBI provides detection results for each window, the proposed RCCA scheme identifies the exact instant of damage through a sharp distortion in its plot.

7.7.3 Damage detection results for 5 DOF B-W system using RCCA

This section deals with the applicability of the proposed RCCA scheme towards damage detection for a nonlinear system. The aforementioned 5 DOF B-W system is considered here to identify both temporal (global) and spatial (local) damage, in an online simultaneous framework. The numerically simulated system is excited with a zero mean Gaussian white noise, sampled at 100 Hz for 50s. The key detection results using the proposed algorithm are presented next in detail.

7.7.3.1 Global damage detection results using the 5 DOF B-W system

This section explores the applicability of the proposed method towards detecting a global damage on the nonlinear 5 DOF B-W system. As excitation to any dynamical system can have white or colored spectra, damage detection in real time using white noise and Kanai-Tajimi (K-T) spectra are considered here. The stochastic K-T model accommodates both time dependent frequency and amplitude modulation. With normalized duration of peak response set at 0.4, the ground motion is simulated for 50s having dominant frequency of earthquake excitation at 3.5 Hz. The global (or temporal) damage is carried out by changing the nonlinear force term κ at 31s by 10%. Detection results obtained using RCCA can be found in Fig. 7.7. The RCCA algorithm proceeds by providing eigenspace

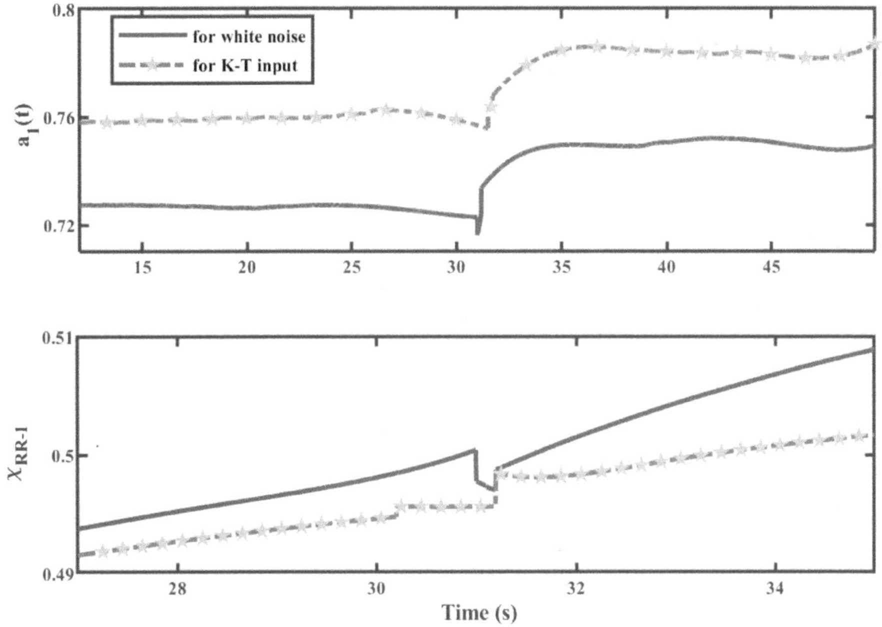

Figure 7.7: Recursive DSF plot for 10% temporal damage for the B-W system (using white noise and K-T spectra).

updates, from which, the transformed responses are obtained at each instant of time. Using TVAR coefficients, the transformed responses are modeled and the detection results using are shown Fig. 7.7. It can be observed from the figure that the plots of the TVAR coefficients clearly identify the damage instant at 31s through a sharp change in the mean level of the plots for both input excitations. In order to confirm the findings, the temporal RRE plot clearly portrays a distortion that confirms the instant of damage at 31s. In line with the above findings, it can be very well established that the proposed RCCA algorithm is efficient in determining fine levels of damage in real time, a feat that is difficult to accomplish, especially when attempted in a recursive framework.

7.7.3.2 Local damage detection results using the 5 DOF B-W system

The applicability of the proposed RCCA algorithm to detect damage in real time for nonlinear systems can be determined through the local damage detection of the aforementioned 5 DOF B-W system. The numerically simulated system is inflicted with a discreet local damage that progresses with the passage of time. The reduction of the storey stiffness of the top floor takes place at 10s from the start of the excitation, that progresses to the fourth floor at 20s, to the third

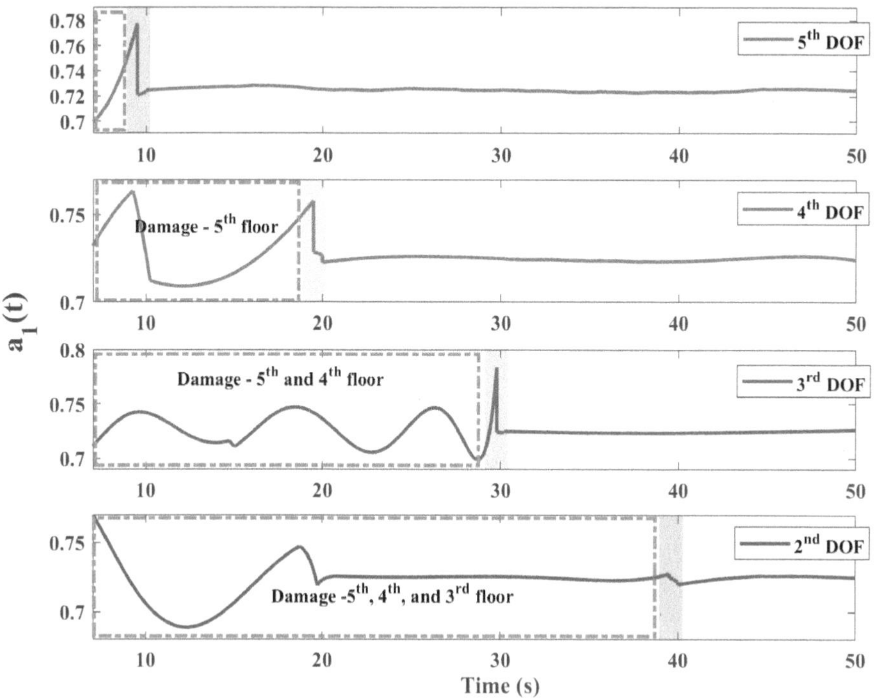

Figure 7.8: Progressive damage to the 5 DOF B-W system.

floor at 30s and eventually proceeds to the second floor at 40s. The base of the system is assumed to remain intact as it is purely nonlinear and plays no part in the reduction of floor stiffness of the structure. A similar phenomenon can be observed in real life situations where the effects of earthquake on a civil structure lead to the deterioration of the building through reduction in storey stiffness. On applying the proposed method on the acceleration responses of the floors, the eigenspace updates at each instant of time are obtained, on which the TVAR models are fit. Figure 7.8 illustrates the progression of damage through the floors, using the proposed method, through the mean shifts of the TVAR coefficients.

It can be observed from the figure that the first reduction of stiffness at 10s is evident from the TVAR plot of the first floor. The absence of any activity after the 10s mark indicates that the damage to the floor has actually occurred at 10s from the start and has progressed on to the next floor. The succeeding plot shows a clear peak at 20s which indicates the commencement of damage to the fourth floor, but the instabilities arising in the pre-damage phase portray the fact that the system has already undergone a certain reduction in stiffness on the previous floor. The culmination of the change in linear stiffness of the second and third floor is purely reflected in the third plot of the same figure, where the distinct peak at 30s is clearly observed. The presence of wavy undulations before the damage instant is reached illustrates the fact that the other floors have already suffered the damage inflicted through a change in stiffness. Finally, in the fourth plot, the peak observed at 40s indicates the exact instant of damage. However, it is observed that there is a lot of activity in the pre-damage zone, the intensity of which is magnified as compared to the other plots. The primary reason behind this phenomenon is that the second floor accounts for the cumulative resultant of all the previously occurred local damage events reflected through the higher amount of instability in the TVAR plots. Therefore, it can be conclusively determined that the proposed RCCA based algorithm can even identify cases of progressive damage, which closely emulates a real life scenario. The recursive implementation of the method enables a continuous monitoring of the system that leads to the simultaneous detection of events that progress over time.

7.8 Experimental verifications

A crucial feature of any algorithm is to assess its effectiveness by validation through experimental case studies, carried out under controlled laboratory environment. In this regard, experimental studies are conducted to demonstrate the applicability of the proposed algorithm. In order to substantiate the robustness of the proposed technique, an experimental validation is presented: A two–story aluminum shear building prototype placed with a water dispenser at the top floor excited by zero mean Gaussian white noise. The experimental setup is excited using dynamic excitations on a shake table. The shake table is specified as: (1) table dimension – 150 cm×150 cm; (2) payload capacity – 5 ton; (3) peak velocity – 153 cm/s; (4) peak acceleration – ±2.0 g; (5) frequency range – 0 – 20 Hz.

7.8.1 2 DOF shear building experimental setup

The aluminum model is $1.5m$ high with inter-storey separation of $0.74m$ fixed on a base plate that is bolted on top of a shake table. The cross section of the base measures $0.3m \times 0.3m \times 0.02m$ placed on the $0.15m$ thick base plate firmly bolted against the shake table. The water dispenser of height $0.26m$ is firmly affixed to the roof plate using commercially available epoxy resin. The model is subjected to scaled zero mean Gaussian white noise excitation and the acceleration data are collected using QuantumX MX410 HBM Data Acquisition (DAQ) system at a sampling frequency of $400Hz$. The shear building model is instrumented at the floor levels in orthogonal directions using PCB Piezotronics triaxial accelerometers of sensitivity $100mV/g$, bearing a frequency range of $0.5 - 10000Hz$. The dispenser when filled with water provides nonlinearity to the setup. This comes into effect when the water sloshes against the walls of the dispenser as the model is subjected to scaled vibration excitation. The flow valve of the dispenser emancipates water at the rate of $1L/30s$ during the run-time of the experiment.

The model is oriented along the axis of the shake table with the local x-axis coinciding with the global x-direction as shown in Fig. 3.14. The experimental trial commences with the water draining out form the dispenser at the rate of $2l/min$ from 20s into the excitation phase. The sloshing of water due to the forced excitation on the prototype accounts for a variable mass loss that forms the basis of transition from a nonlinear to a linear state. From the theoretical understanding of dynamics of vibrating systems, it is a well understood concept that identification of modal parameters of a nonlinear system is difficult and sometimes erroneous. Therefore, a recursive approach employed to identify the modal parameters involves the empty dispenser excited using white noise. The absence of water from the dispenser corresponds to a linear state of the system that leads to the utilization of RCCA as an efficient recursive modal identification tool, the results of which are discussed next in detail.

In order to justify the correctness of the estimated LNMs, an impulse excitation was applied to the top floor of the experimental setup using a rubber mallet and the acceleration response was provided as input to the batch CCA algorithm. The modes obtained by applying batch CCA on the impulse response for the separate set of experimental trials is considered as a ready reference to justify the correctness of estimated LNMs using the RCCA approach. Subsequent analysis using RCCA provides eigenspace updates in real time from which the LNMs can be extracted. It can be understood from Fig. 7.9 that the estimated LNMs tend to converge to the mode shapes obtained from the impulse excitation with the progression of the samples, for both the floor levels. The MAC values recursively obtained for preselected sample points are provided in Table 7.5. Based on the MAC values, it can be well understood that the proposed algorithm provides approximately perfect convergence of the estimated and the mode shapes obtained using impulse excitation. It can now be ascertained that the eigenspace updates reveal useful information regarding the nature of the system at each instant, the utility of which be subsequently extended to detect the transition of state of the system. In this context, the eigenspace updates are tracked recursively on which the TVAR models are fit.

Table 7.5: Iteration vs MAC value for aluminum shear building experiment.

i^{th} **iteration**	MAC_1	MAC_2	i^{th} **iteration**	MAC_1	MAC_2
1000	0.8695	0.8695	13000	0.9901	0.9901
4000	0.9388	0.9388	15000	0.9879	0.9879
6000	0.9632	0.9632	17000	0.9979	0.9979
8000	0.9374	0.9374	19000	0.9985	0.9985
10000	0.9880	0.9880	19300	1.0000	1.0000

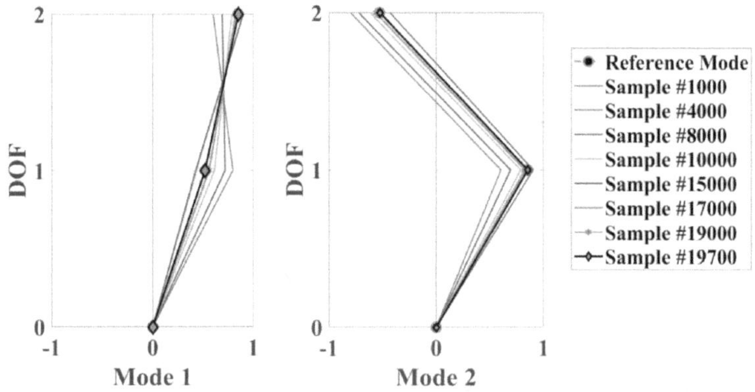

Figure 7.9: Comparison between the modes obtained from impulse response and selected LNMs of 2–DOF shear building model.

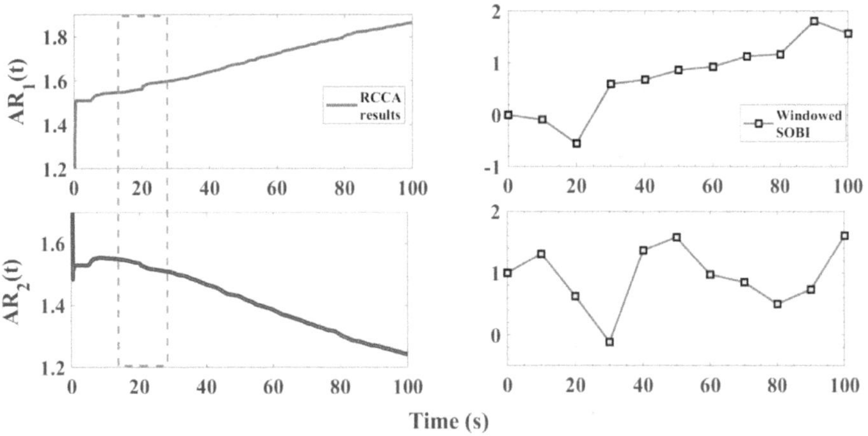

Figure 7.10: DSFs for mass loss at 20s: RCCA vs SOBI.

To provide an insight into the performance of the proposed approach, windowed SOBI method is employed here to detect the transition from nonlinear to linear regime. A comparative study of these methods enables a fair comparison that is significant in establishing the utility of RCCA as an effective real time method for detection. The comparative investigation of these methods is shown in Fig. 7.10. As observed from the figure, the mean shift of the TVAR coefficients commences from 20s that continues until the remainder of the excitation. This phenomenon corresponds to the event of loss of water due to the input excitation and consequent vibration of the model, leading to a variable mass loss. On the contrary, SOBI method provides transformed responses on which windowed AR coefficients at each 10s and 20s intervals are fit. Figure 7.10 clearly illustrates that there is an inconsistent mean change in the level of the AR coefficients. Moreover, there is no significant observation at the 20s mark for both the plots which indicates that the windowed method is not robust enough to identify the change of state. This justifies the necessity of implementing a recursive approach to detect the change of state from a nonlinear to linear regime. Based on the above results, it can be safely concluded that the proposed RCCA method is well adept in identifying both the modal parameters and damage patterns of a system, simultaneously, in real time, in a unified framework, which forms the core essence of carrying out this work.

References

[1] Antoni, J. (2005). Blind separation of vibration components: Principles and demonstrations. Mechanical Systems and Signal Processing, Elsevier, 19(6): 1166–1180.

[2] Cichocki, A. and Amari, S. (2003). Adaptive Blind Signal and Image Processing. John Wiley, New York.

[3] Gul, M. and Catbas, F. N. (2008). Ambient vibration data analysis for structural identification and global condition assessment. American Society of Civil Engineers, 134(8): 650–662.

[4] Brincker, R., Zhang, L. and Anderson, P. (2001). Modal identification of output-only systems using frequency domain decomposition. Smart Materials and Structures, IOP Publishing, 10(3): 441–445.

[5] Sadhu, A., Narasimhan, S. and Antoni, J. (2017). A review of output-only structural mode identification literature employing blind source separation methods. Mechanical Systems and Signal Processing, Elsevier, 94: 415–431.

[6] Antoni, J. and Chauhan, S. (2013). A study and extension of second-order blind source separation to operational modal analysis. Journal of Sound and Vibration, Elsevier, 332(4): 1079–1106.

[7] Rainieri, C. and Fabbrocino, G. (2014). Operational Modal Analysis of Civil Engineering Structures. Springer, New York, 142, 143.

[8] Brincker, R. and Ventura, C. (2015). Introduction to Operational Modal Analysis. John Wiley & Sons.

[9] James, G. H., Carne, T. G., Lauffer, J. P. and others. (1995). The natural excitation technique (NExT) for modal parameter extraction from operating structures. Modal Analysis-the International Journal of Analytical and Experimental Modal Analysis, Blacksburg, VA: Scholarly Communications Project, Virginia Polytechnic Institute & State University, C 1993–1996, 10(4): 260.

[10] Skolnik, D., Lei, Y., Yu, E. and Wallace, J. W. (2006). Identification, model updating, and response prediction of an instrumented 15-story steel-frame building. Earthquake Spectra, 22(3): 781–802.

[11] Juang, J. N. and Pappa, R. S. (1985). An eigensystem realization algorithm for modal parameter identification and model reduction. Journal of Guidance, Control, and Dynamics, 8(5): 620–627.

[12] Amini, F. and Ghasemi, V. (2018). Adaptive modal identification of structures with equivariant adaptive separation via independence approach. Journal of Sound and Vibration, Elsevier, 413: 66–78.

[13] Krishnan, M., Bhowmik, B., Tiwari, A. K. and Hazra, B. (2017). Online damage detection using recursive principal component analysis and recursive condition indicators. Smart Materials and Structures, IOP Publishing, 26(8): 085017.

[14] Krishnan, M., Bhowmik, B., Hazra, B. and Pakrashi, V. (2017). Real time damage detection using recursive principal components and time varying auto-regressive modeling. Mechanical Systems and Signal Processing, Elsevier, 101: 549–574.

[15] Bhowmik, B., Krishnan, M., Hazra, B. and Pakrashi, V. (2018). Real time unified single and multi channel structural damage detection using recursive singular spectrum analysis. Structural Health Monitoring, SAGE Publications, pp. 1475921718760483, doi = 10.1177/1475921718760483.

[16] Hazra, B. (2010). Hybrid Time and Time-Frequency Blind Source Separation Towards Ambient System Identification of Structures. PhD thesis, University of Waterloo.

[17] Allen, M. S., Sracic, M. W., Chauhan, S. and Hansen, M. H. (2011). Output-only modal analysis of linear time-periodic systems with application to wind turbine simulation data. Mechanical Systems and Signal Processing, 25(4): 1174–1191.

[18] Pakrashi, V., O'Connor, A. and Basu, B. (2010). Effect of tuned mass damper on the interaction of a quarter car model with a damaged bridge. Structure and Infrastructure Engineering, 6(4): 409–421.

[19] Hearn, G. and Testa, R. B. (1991). Modal analysis for damage detection in structures. Journal of Structural Engineering, American Society of Civil Engineers, 117(10): 3042–3063.

[20] Hazra, B., Sadhu, A., Roffel, A. J., Paquet, P. E. and Narasimhan, S. (2011). Underdetermined blind identification of structures by using the modified cross-correlation method. Journal of Engineering Mechanics, 138(4): 327–337.

[21] Hazra, B., Roffel, A. J., Narasimhan, S. and Pandey, M. D. (2009). Modified cross-correlation method for the blind identification of structures. Journal of Engineering Mechanics, American Society of Civil Engineers, 136(7): 889–897.

[22] Yang, Y. and Nagarajaiah, S. (2014). Blind identification of damage in time-varying systems using independent component analysis with wavelet transform. Mechanical Systems and Signal Processing, Elsevier, 47(1-2): 3–20.

[23] Hyvärinen, A. and Oja, E. (2000). Independent component analysis: algorithms and applications. Neural Networks, Elsevier, 13(4-5): 411–130.

[24] Feeny, B. F. and Liang, Y. (2003. Interpreting proper orthogonal modes of randomly excited vibration systems. Journal of Sound and Vibration, Elsevier, 265(5): 953–966.

[25] Kerschen, G. and Golinval, J. C. (2002). Physical interpretation of the proper orthogonal modes using the singular value decomposition. Journal of Sound and Vibration, Elsevier, 249(5): 849–865.

[26] Zhou, W. and Chelidze, D. (2007). Blind source separation based vibration mode identification. Mechanical Systems and Signal Processing, Elsevier, 21(8): 3072–3087.

[27] Hazra, B. and Narasimhan, S. (2010). Wavelet-based blind identification of the UCLA Factor building using ambient and earthquake responses. Smart Materials and Structures, IOP Publishing, 19(2): 025005.

[28] Fan, X. and Konold, T. R. (2010). Canonical correlation analysis. Quantitative Methods in the Social and Behavioral Sciences: A Guide for Researchers and Reviewers, pp. 29–34.

[29] Thorndike, R. M. (2000). Canonical correlation analysis. Handbook of Applied Multi-variate Statistics and Mathematical Modeling. Elsevier, pp. 237–263.

[30] Borga, M. (2001). Canonical correlation: A tutorial. On line tutorial http://people. imt. liu. se/magnus/cca. 4(5).

[31] Golub, G. H. and Zha, H. (1994). Perturbation analysis of the canonical correlations of matrix pairs. Linear Algebra and its Applications, 210: 3–28.

[32] Kato, T. (2013). Perturbation theory for linear operators. Springer Science & Business Media, Vol. (132).

[33] Yan, A. M., Kerschen, G., De Boe, P. and Golinval, J. C. (2005). Structural damage diagnosis under varying environmental conditions part II: local PCA for non-linear cases. Mechanical Systems and Signal Processing, Elsevier, 19(4): 865–880.

[34] Watkins, D. S. (1982). Understanding the QR Algorithm. SIAM Review, SIAM, 24(4): 427–440, doi:10.1137/1024100.

[35] Paige, C. C. (1980). Accuracy and effectiveness of the Lanczos algorithm for the symmetric eigenproblem. Linear Algebra and its Applications, 34: 235–258.

[36] Azam, S. E., Chatzi, E., Papadimitriou, C. and Smyth, A. W. (2017). Experimental validation of the Kalman-type filters for online and real time state and input estimation. Journal of Vibration and Control, SAGE Publications, 23(15): 2494–2519.

[37] Li, W., Yue, H. H., Valle-Cervantes, S. and Qin, S.J. (2000). Recursive PCA for adaptive process monitoring. Journal of Process Control, Elsevier, 10(5): 471–48.

[38] Farrar, C. R. and Worden, K. (2007). An introduction to structural health monitoring. Philosophical Transactions of the Royal Society of London A, 365(1851): 303–315.

[39] Salawu, O. S. (1997). Detection of structural damage through changes in frequency: A review. Engineering Structures, Elsevier, 19(9): 718–723.

[40] Pandey, A. K., Biswas, M. and Samman, M. M. (1991). Damage detection from changes in curvature mode shapes. Journal of Sound and Vibration, Elsevier, 145(2): 321–332.

[41] Yang, J. N., Lei, Y., Lin, S. and Huang, N. (2004). Hilbert-Huang based approach for structural damage detection. Journal of Engineering Mechanics, American Society of Civil Engineers, 130(1): 85–95.

[42] Nair, K. K., Kiremidjian, A. S. and Law, K. H. (2006). Time series-based damage detection and localization algorithm with application to the ASCE benchmark structure. Journal of Sound and Vibration, Elsevier, 291(1): 349–368.

[43] Johnson, Erik A., Lam, H. F., Katafygiotis, L. S. and Beck, J. L. (2004). Phase I IASC-ASCE structural health monitoring benchmark problem using simulated data. Journal of Engineering Mechanics, American Society of Civil Engineers, 130(1) 3–15.

[44] Bhowmik, B., Tripura, T., Hazra, B. and Pakrashi, V. (2019). First-order eigen-perturbation techniques for real-time damage detection of vibrating systems: Theory and applications. Applied Mechanics Reviews, 71(6).

[45] Tripura, T., Bhowmik, B., Pakrashi, V. and Hazra, B. (2019). Real-time damage detection of degrading systems. Structural Health Monitoring, p. 1475921719861801.

[46] Caicedo, J. M. (2011). Practical guidelines for the natural excitation technique (NExT) and the eigensystem realization algorithm (ERA) for modal identification using ambient vibration. Experimental Techniques, 35(4): 52–58.

PART II

Chapter 8

Recursive SHM Practical Applications

8.1 Introduction

The robustness of any algorithm can be investigated on its proper applicability on a wide domain of engineering interests – viz., numerical simulations, experimental trials, and practical real-life scenarios. In the interest of providing an insight towards the real-life structural systems monitored using these algorithms, the present chapter provides an executive summary of the monitoring strategies carried out exclusively for the UCLA Factor Building (UCLAFB). The algorithms discussed in the previous chapters will be compared in the context of earthquake identification and prominent damage detection in all the floors of the building. Additionally, the phase-I ASCE-SHM benchmark structure is considered for real-time modal identification studies using RCCA. This chapter promises to be an interesting read for researchers intending to investigate, examine and apply conventional and contemporary algorithms towards monitoring of large infrastructure especially in field studies.

The chapter is organized as follows: First, the applicability of the RPCA-RRE algorithm on the UCLAFB is examined and the results inferred. Next, due to the nonstationary nature of the input, the RPCA-TVAR algorithm is applied on the same dataset obtained from the EW and NS directions of the monitored UCLAFB. With a varying sensor topology, a single sensor is now placed in preselected floors of the UCLAFB to identify the spatio-temporal damage patterns caused due to the Parkfield earthquake. Finally, the ASCE-SHM benchmark time-histories are streamed to identify the vibratory modes in real-time. The results indicate perfect correlation with theoretical structural modes with the progression of samples.

8.2 UCLA factor building—A general overview

The UCLAFB is a 216.5 ft high structure designed and constructed in the late 1970s. The 17 story $G+15$ building (with basement and sub basement levels) comprises special moment resisting steel frames (SMFs) supported by concrete bell caissons and spread footings. The aftermath of the 1994 Northridge earthquake prompted monitoring of the building with a network of 72 Kinemetrics FB-11 uniaxial accelerometers at all the floor levels. The sensors' arrays are further converted to an equivalent set of NS, EW and θ directional readings with each storey having two orthogonal pairs of sensors corresponding to the local coordinates [1–9].

For monitoring the instrumented system, the data is sampled at 100 Hz. In order to identify significant damage patterns and ensure the accurate occurrence of the earthquake, floor acceleration responses due to ambient data and the data recorded during the Parkfield earthquake on September 28, 2004, 10:15 AM PDT (with M= 6.0 on the moment magnitude scale) have been considered [7, 8]. In recent times, UCLAFB has emerged as a leading SHM benchmark that has been extensively reported in reputed journals, case studies and discussed in internationally acclaimed conferences [1–5, 10, 13].

8.3 UCLAFB case study—RPCA-RRE

An extensive database of recorded responses obtained from the monitored UCLAFB is now examined using the RPCA-RRE algorithm. This is the first part of an extensive ambient vibration monitoring program discussed in detail in this chapter.

As envisaged, the roof acceleration response in Fig. 8.1 corresponds to a higher normalized magnitude compared to the fifth floor response [2–4]. It should be noted that this representation provides a subset of a larger database domain and the selection of these vibration signatures are purely arbitrary. Note that a considerable incidence of nonstationary activity is evident around the 380s mark, corresponding to the onset of the Parkfield earthquake. However, this can not be clearly identified due to the extreme nonstationarity of the ground motion. Prior to the occurrence of the earthquake, an ambient vibration regime can be identified from Fig. 8.1.

With real-time data streaming, it was possible to explore the performance of the RPCA-RRE algorithm on the vibration signatures. At each instant of time, these acceleration samples are provided as input to the RPCA-RRE spatio-temporal module, with an aim to identify structural damage patterns online.

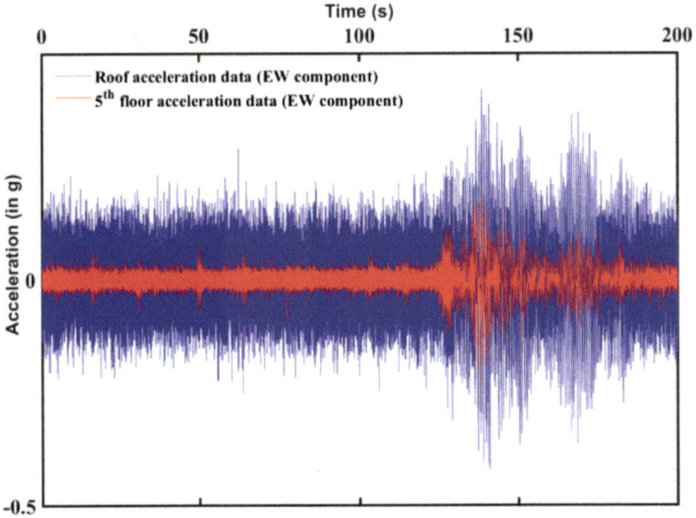

Figure 8.1: EW direction Roof and 5th floor acceleration responses.

Additionally, the commencement of ground shaking (available from the ambient data regime and continued till the end of excitation) could be established from the temporal RRE ($\chi_{RR} - 2$) plots for the $N-S$ and $E-W$ component responses in Fig. 8.2. A sudden shift in the mean level of the plot corresponds to a major damage in both the orthogonal directions. This can be attributed to a closer look at the plot where the system output response shows certain deviations around $380s$, with a pronounced effect at $410s$ – which is evident from the RPCA-RRE analysis (Fig. 8.2). From an engineering point of view, it is natural to expect these two time stamps to differ by a margin as the vibrating system requires a certain amount of time to indicate a significant distortion in its parameters (usually evident from the changes in stiffness of the floor levels).

Previous studies carried out by this research group have reported the following set of frequency reduction percentages between ambient vibration and earthquake data [1]: 14.31%, 13.95%, 15.61%, 8.49%, 7.24%, 4.95%, 4.95%,30%, 6.5%, 4.75%. This corroborates to a significant global reduction in structural stiffness between the columns of floor levels. The focus of the present work uses a percentage change in the average spatial RRE values (i.e., $\Delta\langle \varepsilon_{RR} - Y_i \rangle$ using Eq. 4.19) between the ambient and earthquake regimes corresponding to pre and post damage scenarios. Corresponding to pre-selected floor levels, Fig. 8.3 presents the distortions in spatial RRE patterns $\varepsilon_{RR} - Y_i$. Evidently, the spatial RRE plot for the 3rd, 8th and 15th floors demonstrate moderate shifts in the mean level at around t=$380s$ and significant changes in the neighbourhood of t=$410s$ indicating

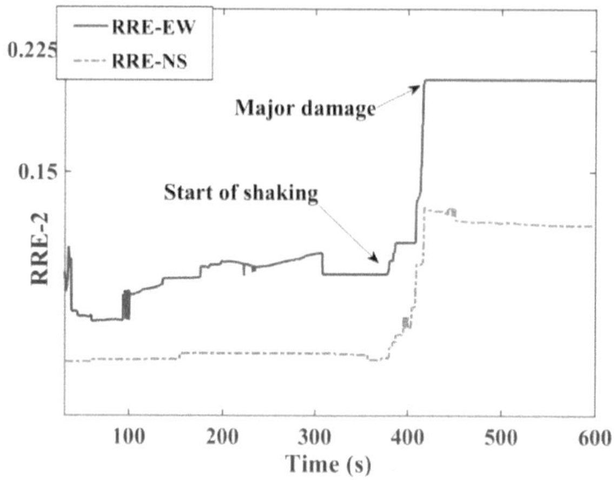

Figure 8.2: Residual error plots for UCLA in EW and NS directions.

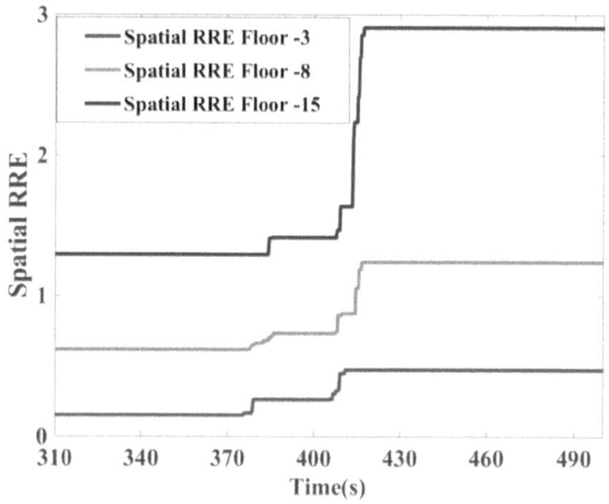

Figure 8.3: Residual error plots for floors of UCLA.

the occurrence of damage. The percentage change in post damage RRE and pre damage RRE (i.e., $\Delta\langle \varepsilon_{RR} - Y_i \rangle$) for each floor as shown in Table 8.1 indicates the appearance of damage not only at a single floor but the system as a whole which corroborates to the previously mentioned results on modal identification [1].

The applicability of the RPCA-RRE algorithm to identify damage patterns for under determined systems has already been discussed at length in the previous chapters. To ensure robustness of the devloped approach, it becomes essential to engineer a case-specific situation where the instrumented DOFs are lower than

Table 8.1: Spatial RREs for the UCLA factor building.

	Y-EW			Y-NS		
Response No	Pre-damage	Post-damage	$\Delta\varepsilon_{RRE}$(%)	Pre-damage	Post-damage	$\Delta\varepsilon_{RRE}$(%)
1	2.03	14.76	86.25	3.86	17.84	78.34
2	3.70	40.90	90.95	6.34	28.09	77.42
3	4.93	37.46	86.84	3.45	34.43	89.98
4	4.43	28.26	84.32	4.54	13.05	65.24
5	2.22	21.45	89.63	1.19	10.77	88.97
6	1.93	27.68	93.04	2.12	26.04	91.85
7	1.84	19.12	90.37	2.01	12.34	83.74
8	1.79	63.84	97.20	1.74	15.21	88.55
9	1.52	41.65	96.34	1.55	24.29	93.63
10	3.08	21.08	85.39	6.96	50.89	86.33
11	2.14	49.98	95.72	1.39	40.71	96.58
12	2.01	79.54	97.47	3.60	31.07	88.42
13	2.01	52.56	96.17	1.59	28.09	94.36
14	3.14	86.25	96.36	2.70	21.94	87.70
15	50.10	240.64	79.18	6.18	61.10	89.88
16	31.00	69.27	55.24	11.75	104.82	88.79

the actual number of DOFs – meaning, a subset of system responses is to be used for fault identification. Considering an array of all the odd DOFs (viz., 1st, 3rd, ..., 15th floor responses), this subset of responses is provided as an input to the RPCA-RRE framework. Figure 8.4 clearly demonstrates that χ_{RR-2} accurately identifies the instant of damage. This strengthens our belief in the practical applicability of the RPCA-RRE method in determining events of interest for large scale real-life systems.

Certain downsides to the RPCA-RRE algorithm in the context of practical applicability are summarized below. Our experience here should be treated as pitfalls to be avoided while handling actual building structures with significantly higher DOFs.

1. On the downside, the RPCA-RRE method needs improvement in dealing with the non-stationarity of input data. In situations where ground motion is the primary excitation source, masking of damage sensitive features is evident from the plots shown above.

2. For practical cases, systems are generally time-varying. In order to capture the essence of detection in a real-time framework, it becomes necessary to model the system in a time-varying framework, which will be demonstrated in the following sections.

3. For under determined systems, a large subset of sensor information (around 50%) is necessary to be provided as input to the RPCA-RRE algorithm.

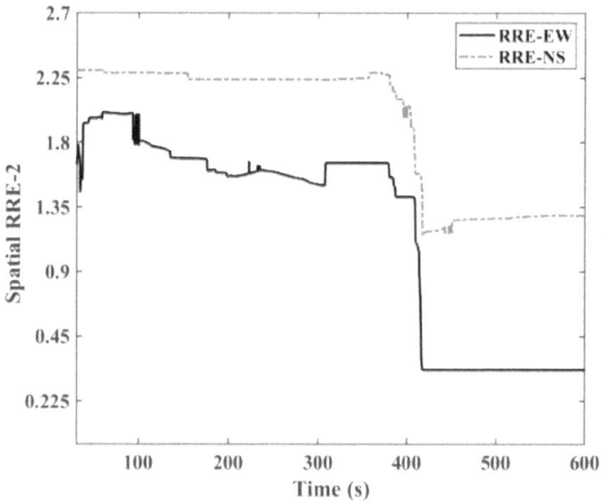

Figure 8.4: Residual error plots for UCLA underdetermined case.

Cases where instrumented DOFs are significantly lower than the actual DOFs (10% or less), it becomes difficult for the method to accurately identify the damage instant and pinpoint the location of damage, in a single online framework.

4. Although the present algorithm achieved significant success in detecting spatial and temporal damage simultaneously in simulation scenarios, the combined presence of amplitude and frequency domain nonstationarities in excitation still poses challenges in simultaneous spatial and temporal detection of damage especially in field implementations.

An attempt to overcome the shortcomings stated above is presented in detail from the next section onwards.

8.4 UCLAFB case study—RPCA-TVAR

An important aspect to be ventured into while dealing with damage detection strategies is its applicability in identifying events that evolve from real life scenarios. In the previous section, a nifty illustration for the same was proposed using the ambient and recorded earthquake vibration data collected from the heavily instrumented UCLAFB. In the present context, the same exemplar is revisited, primarily to perceive the key differences and understand the subtle nuances that arise due to the implementation of the newer proposed methodology based on the RPCA-TVAR approach [2, 11]. In the later stages of this chapter, a comparative study of all the FOEP based techniques would be portrayed, reviewing the

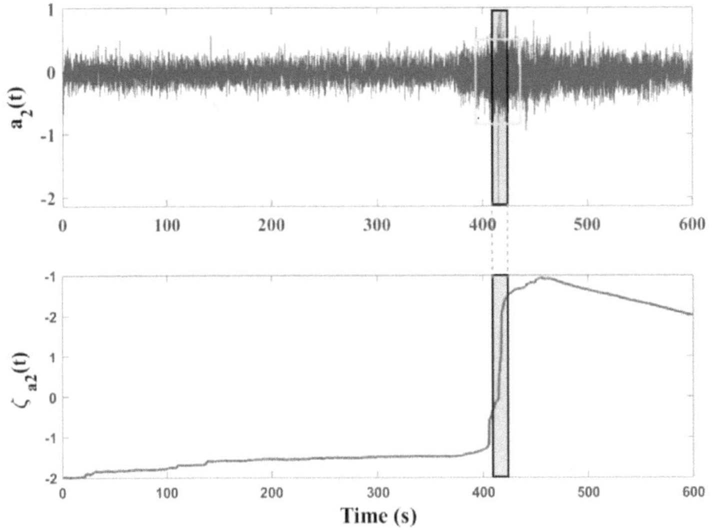

Figure 8.5: DSF for the UCLAFB using RPCA-TVAR.

featured detection results based on the input vibration data of the same model. Towards this, a solid build-up for the comparative study using detection results of various algorithms on the UCLAFB recorded vibration data is an important aspect that needs to be addressed.

The RPCA-TVAR method has been previously discussed in detail to demonstrate its effectiveness in situations where systems are time-varying. For the present case, the recorded acceleration data from the monitored sensors of the UCLAFB is provided as an input to the real-time algorithm. The primary report revealed that the transformed responses are inadequate by themselves to extract the instant of damage. Subsequently, the natural flow of the algorithm initiated the DSFs to be automatically applied on the transformed response. Recall that the temporal RRE plots illustrated in the previous chapter (Fig. 8.2), failed to portray significant deviations at the onset of the earthquake at 380s but could clearly capture the prominent damage instant at around the 410s mark. Evident from Fig. 8.5, the TVAR coefficient shows a significant amount of activity around the 410s mark, which has been highlighted. Contrary to the RREs (discussed in the previous section), the HOM applied on the TVAR coefficient clearly depicts the onset of the earthquake as well as the start of the significant damage around the vicinity of the 410s time stamp. This can be attributed visually to the change in the mean level of the plot.

Recall that the previous discussion on changes in local damage attributes was structured in a tabular format. The present context illustrates this aspect in a

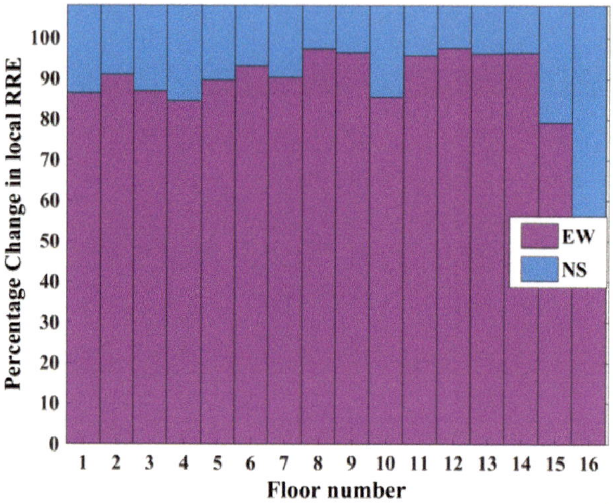

Figure 8.6: Bar diagram indicating percentage change in local RRE for EW and NS direction.

graphical format provided in Fig. 8.6. This local damage is expressed through a percentage change in the average spatial RRE values (i.e., $\Delta\langle \varepsilon_{RR} - Y_i \rangle$ between the ambient and earthquake regimes corresponding to pre and post damage scenarios. The plot indicates the appearance of a global damage which is manifested in the form of individual and coupled local damages in the representative floor levels. This corroborates to the previously mentioned results on modal identification [1,9].

Prior experience in this area reveals specific pitfalls in the implementation of the RPCA-TVAR algorithm. These can be summarized below for reference:

1. Although a predominant part of DSF masking has been resolved using RPCA-TVAR, it is a firm belief of the authors that an additional online filtering scheme is necessary to eradicate this issue completely. Situations (such as this) which require the use of online filtering techniques can be addressed using the RSSA algorithm. The performance study on the UCLAFB is discussed in the following sections.

2. For practical situations, cost, inaccessibility and the unavailability of good quality sensors impede the functioning of a complete monitoring setup. These situations require the use of a single sensor that can be mobilized over a pre-selected network topology to extract meaningful information regarding system health. Multi-channel sensing algorithms such as the RPCA-RRE and the RPCA-TVAR are not mathematically consistent with single sensor topologies and therefore, require the use of a RSSA-based single sensor framework.

3. A generic monitored building is never expected to be excited with ground motion *perfectly* aligned to its local orientation. For a more extensive practical corroboration of the UCLAFB events, bi-directional dependencies are considered in the following sections through the use of RCCA-based methodologies. As previously discussed, a seamless transition from field to practice is made possible with the use of RCCA; discussed in detail in the following sections.

8.5 UCLAFB case study—RSSA

Practical considerations, shortage of funds, passiveness in compliance, and inaccessibility are some major concerns that impede proper SHM monitoring of built infrastructure. These circumstances in particular, necessitate the installation of a single (good) quality sensor for monitoring. Now, the mobility of such a sensor – typically an accelerometer – is not compromised due to its portability. This can ensure a systematic network topology where the movable sensor is placed at the DOF of interest to identify the spatial and temporal patterns of damage [3, 5]. Similar to this train of thought, the UCLAFB is considered to be instrumented on a pre-selected DOF for real-time detection of system damage due to the Parkfield earthquake.

Out of the 48 responses obtained from the data set, the translational responses are utilized for online processing using the proposed algorithm. First, the acceleration response of the roof corresponding to the E-W component is provided as an input to the RSSA module. Next, an arbitrary floor level (3rd floor in this case) is chosen for analysis. TVAR modeling on the set of reduced order responses from RSSA is carried out on both the time histories. Additionally, recursive eigen ratio difference (ERD) corresponding to both the reduced order responses are evaluated. A graphical representation of both the DSFs is presented in Fig. 8.7.

The salient points of interest inferred from the graph validating the strong suit of the algorithm are as follows:

1. Even with a nonstationary input excitation, the TVAR coefficients and the recursive ERD plots are distinctly represented. This reinforces the *real-time filtering property* of the RSSA algorithm which automatically removes instances of masking as was previously evident with RPCA-based methods.

2. Around the 380s mark, well-defined recognizable patterns can be visualized from both the DSF plots. As previously discussed, this time stamp corresponds to the start of the shaking event – detected by both RPCA-based algorithms – which is significantly apparent with the use of RSSA as well. Contrary to the previous instances using RPCA (Figs. 8.2 and 8.5), where the instant of shaking is less evident compared to significant damage

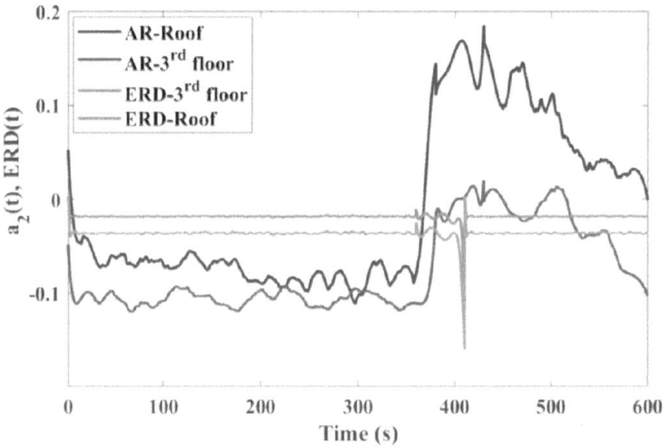

Figure 8.7: DSFs for the UCLAFB using RSSA.

event, the RSSA algorithm provides a sudden mean shift that enhances the occurrence of the ground shaking.

3. Now that the TVAR coefficient clearly indicates the instant of shaking, it becomes important to identify the pronounced damage effect on the system. Manifested usually in the form of stiffness changes (both globally and incrementally, locally), the ERD examined in a recursive framework indicates visibly recognizable spikes around the 410s time stamp. Evident from the previous discussions, the RSSA algorithms therefore, effectively identify the damage event properly utilizing single sensor information.

Hence, it can be concluded that the proposed algorithm could be extensively used for damage detection of real life structures involving large data dimensions under nonstationary cases as well.

8.6 Case study for the IASC–ASCE benchmark structure

For a complete understanding of real-time SHM, practical case-specific scenarios should include modal identification investigations in addition to damage detection. RCCA, a prominent algorithm in this regard, has been extensively utilized for real-time modal identification studies towards a practical situation. In this context, the widely-popular phase-I ASCE-SHM benchmark structure is considered for online analysis.

8.7 Phase-I ASCE-SHM benchmark—A general overview

The phase I–International Association of Structural Control–American Society of Civil Engineers (IASC–ASCE) [12–14] SHM benchmark problem is considered where the proposed algorithm is employed for identification of the vibration modes of the structure in real time. The IASC–ASCE structure is a 3.6m tall, four-storey, two–bay by two–bay steel frame with a 2.5m square base. The benchmark steel frame is a quarter–scale structure created by IASC–ASCE Structural Health Monitoring Task Group located at University of British Colombia, widely used for validating SHM benchmark problems. Interested readers are referred to [12–14] for more details about material and sectional properties of the structural members and definition of the test structure. As the key objective of this section is to demonstrate the utility of the RCCA algorithm in identifying vibration modes of the system in real time, the pristine (undamaged) state of the benchmark corresponding to ID–0 (default case) is considered. The 12–DOF shear–building model is assumed to have 3–DOF per floor level, corresponding to the translation in the x and y directions and rotation θ about the center column. The floors are considered as rigid bodies and the system mass and stiffness details along with the possible damage patterns can be obtained from [12]. For ease of analysis, the building is simplified as a 4–DOF symmetric system with a load at all storeys. To generate simulated vibration data the excitation is modeled as Gaussian white noise and applied to each floor slab in the positive y (weak) direction. For the present case, the generated white noise excitation has a root mean square (RMS) value of 10% of the largest RMS of the acceleration responses (typically of the roof responses). A damping of 1% is introduced in the system (default), and the white noise is sampled at 500Hz for a reasonable record length of 100s. The identification of vibrating modes commences at the onset of the streaming data. At each instant of time, the recursive modal identification module identifies the vibrating modes of the system, represented in Fig. 8.8. The consistent evolution of MAC values in a recursive fashion indicates that the LNMs converge to the theoretical vibrating modes at approximately 17%–18% (when the MAC$\geqslant$0.95) of the entire sample population. For simplicity, the recursive MAC values corresponding to pre–selected sample points are provided in Table 8.2. From the table, it can be inferred that the recursively estimated MAC values for each of the modes steadily approaches to approximately 0.98, indicating a excellent correlation of the LNMs to the theoretical modes. The plots of the theoretical modes corresponding to the pre–selected sample points are shown in Fig. 8.9. It can be easily observed that the recursive estimation of the modes are reasonably accurate, thereby indicating the efficacy of the proposed real time modal identification approach towards practical scenario as well.

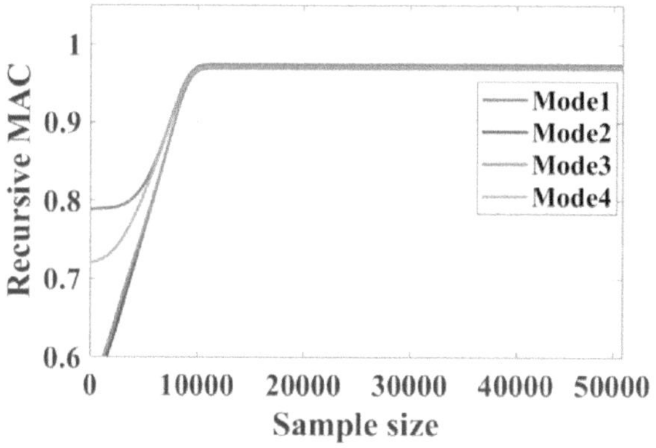

Figure 8.8: Recursive evolution of MAC values for the IASC–ASCE benchmark problem.

Table 8.2: Recursive MAC value for IASC–ASCE benchmark structure.

i^{th} **iteration**	MAC_1	MAC_2	MAC_3	MAC_4
1000	0.7774	0.5743	0.5935	0.7311
5000	0.7815	0.6563	0.6652	0.7512
10000	0.9737	0.9669	0.9699	0.9701
20000	0.9712	0.9666	0.9693	0.9714
30000	0.9727	0.9693	0.9691	0.9724
40000	0.9751	0.9665	0.9679	0.9729
45000	0.9736	0.9624	0.9719	0.9701
50000	0.9752	0.9622	0.9738	0.9729

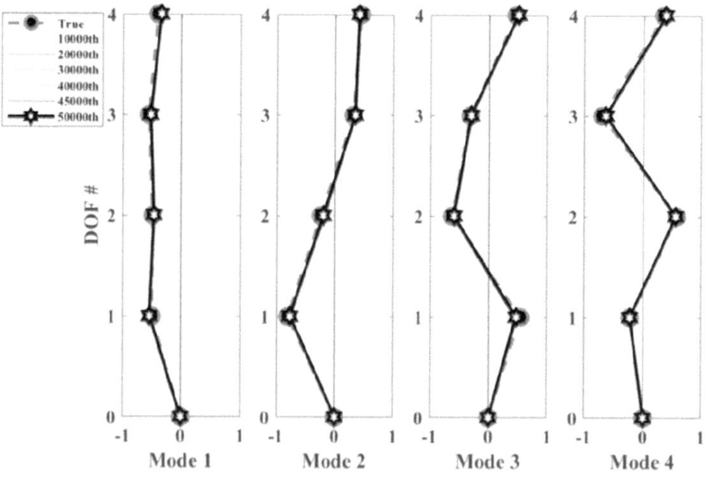

Figure 8.9: Comparison between the theoretical modes and LNMs for the IASC–ASCE benchmark problem.

References

[1] Hazra, B. and Narasimhan, S. (2010). Wavelet-based blind identification of the UCLA Factor building using ambient and earthquake responses. Smart Materials and Structures, IOP Publishing, 19(2): 025005.

[2] Krishnan, M., Bhowmik, B., Tiwari, A. K. and Hazra, B. (2017). Online damage detection using recursive principal component analysis and recursive condition indicators. Smart Materials and Structures, IOP Publishing, 26(8): 085017.

[3] Krishnan, M., Bhowmik, B., Hazra, B. and Pakrashi, V. (2017). Real time damage detection using recursive principal components and time varying auto-regressive modeling. Mechanical Systems and Signal Processing, Elsevier, 101: 549–574.

[4] Bhowmik, B., Krishnan, M., Hazra, B. and Pakrashi, V. (2018). Real time unified single and multi channel structural damage detection using recursive singular spectrum analysis. Structural Health Monitoring, SAGE Publications, pp. 1475921718760483, doi = 10.1177/1475921718760483.

[5] Bhowmik, B. (2018). Online structural damage detection using first order eigenperturbation techniques (Doctoral dissertation).

[6] Bhowmik, B., Tripura, T., Hazra, B. and Pakrashi, V. (2019). First-order eigen-perturbation techniques for real-time damage detection of vibrating systems: Theory and applications. Applied Mechanics Reviews, 71(6).

[7] Kohler, M. D., Davis, P. M. and Safak, E. (2005). Earthquake and ambient vibration monitoring of the steel-frame UCLA Factor building. Earthquake Spectra, 21(3:, 715–736.

[8] Baek, J. H., Hansen, M., Nigbor, R. and Tileylioglu, S. (2006, July). Elevators as an excitation source for structural health monitoring in buildings. In Proceedings of the 4th World Conference on Structural Control and Monitoring, Los Angeles, CA, USA, pp. 11–12.

[9] Bhowmik, B., Tripura, T., Hazra, B. and Pakrashi, V. (2020). Real time structural modal identification using recursive canonical correlation analysis and application towards online structural damage detection. Journal of Sound and Vibration, 468: 115101.

[10] Nayeri, R. D., Masri, S. F., Ghanem, R. G. and Nigbor, R. L. (2008). A novel approach for the structural identification and monitoring of a full-scale 17-story building based on ambient vibration measurements. Smart Materials and Structures, 17(2): 025006.

[11] Bhowmik, B., Tripura, T., Hazra, B. and Pakrashi, V. (2020). Robust linear and nonlinear structural damage detection using recursive canonical correlation analysis. Mechanical Systems and Signal Processing, 136: 106499.

[12] Johnson, E. A., Lam, H. F., Katafygiotis, L. S. and Beck, J. L. (2004). Phase I IASC-ASCE structural health monitoring benchmark problem using simulated data. Journal of Engineering Mechanics, 130(1): 3–15.

[13] Johnson, E. A., Lam, H. F., Katafygiotis, L. S. and Beck, J. L. (2001). A benchmark problem for structural health monitoring and damage detection. In Structural Control for Civil and Infrastructure Engineering, pp. 317–324.

[14] Lam, H. F., Katafygiotis, L. S. and Mickleborough, N. C. (2004). Application of a statistical model updating approach on phase I of the IASC-ASCE structural health monitoring benchmark study. Journal of Engineering Mechanics, 130(1): 34–48.

Chapter 9

Conclusions and Future Prospects

9.1 Summary

Overall, the book provides an approach towards real-time detection of built infrastructure systems, which can be translated and adapted to the performance criteria of several other sectors with minimal change. In that sense, this book presents a paradigm of real-time monitoring and assessment of dynamical systems, with a focus on the built infrastructure sector. However, the need for a critical understanding of each sector and its detection requirements will remain, and this cannot be replaced with algorithms or technology.

Initially, the context of real-time detection, especially around the evolution of the topic of Structural Health Monitoring and damage detection is presented with some core conceptual ideas which form the building blocks of the subsequent chapters. Subsequently, real-time detection with a single sensor is focused on, since this is often the case for many structures. This approach also highlights the value of obtaining such information with minimum instrumentation. Detection is considered for both spatial and temporal domains. The idea is then established in detail for real-time damage detection through a multi-sensor framework and extensive numerical and experimental examples. Subsequently, this idea is extended for real-time modal identification and its implementation is presented for large scale benchmark systems.

9.2 Overview of Results and Contributions

Chapter 3 provides a framework for single-sensor damage detectionin real-time. The RSSA has shown to aid in noise removal online that alleviates the need for implementing traditional filters for pre-processing. Good detection performance in operational conditions and for non-stationary excitation such as earthquakes, with successful detection levels as low as for 15% stiffness change for linear and nonlinear conditions have also been successfully demonstrated in the chapter.

A Buoc-Wen model is used to illustrate multi-sensor damage detection techniques with RPCA. Through **Chapter 4**, recursive variation of the traditional principal component analysis based on the fundamentals of FOEP provides a numerical investigation of spatio-temporal damage detection for 25% damage. Case studies with underdetermined systems additionally indicate the efficacy of RPCA in adjusting to more practical situations where instrumentation cost precedes the value of monitoring.

The ideas of multi-sensor damage detection evolve into a more generalized framework where time-varying auto-regressive models are used to identify patterns in damage and their detection in spatial and temporal domains. The fundamentals of TVAR modelling are presented next in **Chapter 5** and are highlighted through real-time damage detection studies of 15% changes in nonlinearities. Results for El Centro ground excitation and a very specific case of time-diluted degradation form the highlights of this chapter.

In **Chapter 6**, real-time spatio-temporal damage detection case studies are provided using RCCA. This online algorithm employs bi-directional capabilities and accommodates input excitations from multiple directions that excite a dynamical system. Finer damage detection results as low as 10%, on comparison with RPCA for the Buoc-Wen model, and detectability for a strongly nonlinear Duffing Oscillator call attention to the applicability, scaling, and robustness of the RCCA approach.

One of the key characteristics of a SHM module is to identify the dynamic modes of vibration of a monitored system. Towards this, **Chapter 7** provides a comprehensive evidence base in employing RCCA as a *real-time* modal identification algorithm. Evolution of linear normal modes provides a ground for comparison with the theoretical eigenvectors using MAC values. The convergence of MAC values to 1 (unity) from 2-% of an ensemble population builds our confidence in the applicability of RCCA as an online modal identification tool.

SHM enthusiasts will regard **Chapter 8** as one of the central aspects of real-time monitoring. The previous chapter that you have read provides a summary of all the discussed real-time FOEP-based algorithms in light of practical investigations. Specifically, monitoring data from the UCLA Factor Building during the Parkfield Earthquake was used to provide insights into the damage instant and the pronounced effect of the earthquake. Convincingly, RSSA and RCCA have shown to outperform RPCA—mainly due to the online filtering property (of RSSA) and the intelligible bi-directional capabilities (for RCCA)— towards distinguishing the start of ground excitation (at 380s) and its pronounced effect (410s). The findings have firmly led credence to the fact that FOEP-based recursive real-time approaches discussed in this book hold promise towards integration with hardware for *in-situ* practical SHM activities.

9.3 Future Prospects of Real-Time Monitoring

A number of aspects are expected to be of importance in the evolution of real-time monitoring, and several of them are overall related to how SHM will evolve as well.

With the significant increase in the numbers and types of sensors available, possibilities of measurement have increased very significantly over the last decade [3, 10, 16] and terms like Internet of Things (IoT) or are becoming a part of the common parlance in this field. However, this has also raised an increased need to understand the quality of data that is measured, along with the context of the measurement and the detection targets. Decisions on measurement locations, designing data collection and a monitoring framework, understanding the fusion of data coming from multiple sensors have become significant challenges. Additionally, there are also debates around whether several cheap sensors can be as good or better than a few, carefully placed high fidelity sensors. Discussions around smartphone-driven assessments and citizen science often converge to this question. There are no straightforward overall answers to these but several used cases and benchmarks can provide best practice guidelines and limits of applicability and interpretation.

A wider and better understanding of measurements are often required, especially around the calibration and uncertainties of sensors deployed in such real-time monitoring solutions. There may be a range of sources of such monitoring and there can also be uncertainties around the reporting of biases or errors in measurement. They also include questions around missing data and their imputation [2], presence of slow and fast features in data [7], mismatch of time stamps in data and non-uniform sampling, low sampling and the presence of noise in its widest sense. A better control over the uncertainties fundamental to the measurements, or from numerical operations of measurement, or even from assimilating various sources of data can be of significant relevance and importance. Both qualitative and quantitative evidence base around this challenge is important, but a probabilistic format for such investigations can be particularly helpful.

Real-time detection consists of methods and markers that relate to the feature of interest in the structure. While comparisons are often present in published literature for various methods, a significant number of them tend to have an objective of establishing a proposed method to be superior than other competing methods. This is often a problem as it relates back to the prioritisation of the best method above others, whereas in real situations, a range of methods tend to provide similar levels of performance, with one method prevailing over another in terms of a performance or error metric. An alternative viewpoint is becoming popular, which the book has tried to adhere to, where the definition and method of comparison is often the focal point rather than trying to establish or find the very best method. This approach creates sets of allowable approaches and indicators with overall similar levels of performance, but establishes the levels of errors, their definitions and metrics very clearly. It is expected that in future this robust understanding and interpretation of errors and performance will be followed, rather than trying to obtain a slight gain through a bespoke method based on a certain definition of performance. This will also create a better set of deployment conditions with realistic expectations of performance from real-time monitoring.

The levels of performance, uncertainty and even the ability of a certain method to pick up a feature of interest is related to the specific needs of a sector. It also dictates the choice of sensor [8], measurement regime, location of sensors and the interpretation of data quite often. While this relationship is well-accepted, there is still a paucity in literature in establishing a sector specific evidence base [14], both qualitatively and quantitatively, for real-time monitoring. Such benchmarks must be created and it will also provide a better understanding of not only what real-time monitoring can do, but also where the limitations are and what performances or operational regimes are not trustworthy for a certain problem. This is often an important requirement from the standpoint of the industry that is overlooked. This will also impact the future evolution of digital twinning.

The measurement and monitoring of such systems are often carried out at locations where communication of data can be a challenge. While data communication has improved very significantly, an ideal option will be to carry out as much analyses as possible at the node where the data is being measured. Minimisation of energy requirement and the consideration of energy generated from harvesting [1, 15] has also been discussed around such a local computing framework, as has been the possibility of minimizing latency of data [12]. In this context, the evolution of edge computing has been strong in recent years [5] and this can lead to a significant breakthrough in the challenge around data communication and computing, by adapting and developing a local or distributed framework as opposed to its centralized counterpart. Such an approach can also address issues around big data to a significant extent and retain only the consistent detectors [6] at a frequency that is relevant to decision making.

While data is central to the development of these analytics, the interpretation and decision-making is unlikely to be ever completely automatic. The problems and industry specific interpretations are often unique and so while automation will increase, it is unlikely that we will ever make human interaction completely redundant. Under such circumstances, it is important to unify or make semantics [11] around such real-time detections frameworks as consistent as possible. It will allow for better interpretation, comparison and also future development of solutions.

The Value of Information (VoI) [4, 9] is a core aspect around decision making for the users and owners of the data and a consistent framework will also allow us to estimate VoI from real-time monitoring, which in turn qualitatively and quantitatively informs us on the requirement, choice and impact of real-time monitoring, along with inspections [13]. Not only will it let us choose the right scenarios for real-time monitoring, but also provide a better rationale and quantified impact of such monitoring for a certain sector. This can help significant decision-making in terms of commercial or operational decisions, along with decisions on rehabilitation, repair, or rebuilding.

Real-time monitoring is evolving fast but requires normative guidelines to be a part of wider safety frameworks in different sectors. In this regard, standardization is of particular importance. While the process of standardization can be slow, evidence bases presented here and proposed for the future can collectively forge a pathway towards standardization, which will subsequently allow a wider range of commercial entities to adapt it more readily and for a wider range of application areas.

9.4 Conclusions

Structural Health Monitoring (SHM) has been evolving since its inception, and quite dynamically so. There remain extensive opportunities to integrate our fundamental understanding around the real-time detection needs of various systems and data obtained from a wide range of sensors that are being created. On the other hand, the context, interpretation, performance and challenges of real-time detection poses a wide range of challenges. In future, we expect to see a wider range of such challenges, their contextual interpretation, need to adapt to domain knowledge and a wider range of real-time techniques and development of features of interest. It will remain important to investigate such needs and solutions, along with their performances from a qualitative and quantitative sense. Comparison of such performances should be carried out carefully and there is a need to consider a set of admissible methods relevant for a detection need, rather than trying to search for the best method, based on a specific performance criteria. This will make real-time detection holistic and expectation of their performance more realistic when implemented for problems in different sectors.

References

[1] Ali, S. F., Friswell, M. I. and Adhikari, S. (2011). Analysis of energy harvesters for highway bridges. Journal of Intelligent Material Systems and Structures, 22: 1929–38.

[2] Buckley, Tadhg, Vikram Pakrashi and Bidisha Ghosh. (2021). A dynamic harmonic regression approach for bridge structural health monitoring. Structural Health Monitoring.

[3] Zanella, Andrea, Nicola Bui, Angelo Castellani, Lorenzo Vangelista et al. (2014). Internet of Things for smart cities. IEEE Internet of Things Journal, 1: 22–32.

[4] Kundu, Tribikram, Matteo Pozzi and Armen Der Kiureghian. (2011). Assessing the value of information for long-term structural health monitoring. In Health Monitoring of Structural and Biological Systems 2011.

[5] Guo, W., Fouda, M. E., Eltawil, A. M. and Salama, K. N. (2021). Neural coding in spiking neural networks: A comparative study for robust neuromorphic systems. Front Neurosci., 15: 638474.

[6] Klis, Roman and Eleni N. Chatzi. (2016). Vibration monitoring via spectro-temporal compressive sensing for wireless sensor networks. Structure and Infrastructure Engineering, 13: 195–209.

[7] Lizarraga, Ian and Martin Wechselberger. (2020). Computational singular perturbation method for nonstandard slow-fast systems. SIAM Journal on Applied Dynamical Systems, 19: 994–1028.

[8] Lynch, J. P. (2007). An overview of wireless structural health monitoring for civil structures. Philos Trans A Math Phys Eng Sci, 365: 345–72.

[9] Malings, C. and Pozzi, M. (2018). Value-of-information in spatio-temporal systems: Sensor placement and scheduling. Reliability Engineering & System Safety, 172: 45–57.

[10] Mekki, Kais, Eddy Bajic, Frederic Chaxel, Fernand Meyer et al. (2019). A comparative study of LPWAN technologies for large-scale IoT deployment. ICT Express, 5: 1–7.

[11] O'Byrne, Michael, Vikram Pakrashi, Franck Schoefs, Bidisha Ghosh et al. (2018). Semantic segmentation of underwater imagery using deep networks trained on synthetic imagery. Journal of Marine Science and Engineering, 6.

[12] Piyare, R., Murphy, A. L., Magno, M. and Benini, L. (2018). On-demand LoRa: Asynchronous TDMA for energy efficient and low latency communication in IoT. Sensors (Basel), 18.

[13] Quirk, Lucy, Jose Matos, Jimmy Murphy, Vikram Pakrashi et al. (2017). Visual inspection and bridge management. Structure and Infrastructure Engineering, 14: 320–32.

[14] Sohn, H. (2007). Effects of environmental and operational variability on structural health monitoring. Philos Trans A Math Phys Eng Sci, 365: 539–60.

[15] Srbinovski, B., Magno, M., Edwards-Murphy, F., Pakrashi, V., Popovici, E. et al. (2016). An energy aware adaptive sampling algorithm for energy harvesting WSN with energy hungry sensors. Sensors (Basel), 16: 448.

[16] Di Pascale, Emanuele, Irene Macaluso, Avishek Nag, Mark Kelly et al. (2018). The network as a computer: a framework for distributed computing over IoT mesh networks. IEEE Internet of Things Journal, 5: 2107–19.

Index

O

P

Q

R

S

T